The Fundamentals of Quality Assurance in the Textile Industry

The Fundamentals of Quality Assurance in the Textile Industry

Stanley Bernard Brahams

CRC Press
Taylor & Francis Group
Boca Raton London New York

CRC Press is an imprint of the
Taylor & Francis Group, an **informa** business

A PRODUCTIVITY PRESS BOOK

CRC Press
Taylor & Francis Group
6000 Broken Sound Parkway NW, Suite 300
Boca Raton, FL 33487-2742

© 2017 by Taylor & Francis Group, LLC
CRC Press is an imprint of Taylor & Francis Group, an Informa business

No claim to original U.S. Government works

Printed on acid-free paper
Version Date: 20160912

International Standard Book Number-13: 978-1-4987-7788-9 (Hardback)

This book contains information obtained from authentic and highly regarded sources. Reasonable efforts have been made to publish reliable data and information, but the author and publisher cannot assume responsibility for the validity of all materials or the consequences of their use. The authors and publishers have attempted to trace the copyright holders of all material reproduced in this publication and apologize to copyright holders if permission to publish in this form has not been obtained. If any copyright material has not been acknowledged please write and let us know so we may rectify in any future reprint.

Except as permitted under U.S. Copyright Law, no part of this book may be reprinted, reproduced, transmitted, or utilized in any form by any electronic, mechanical, or other means, now known or hereafter invented, including photocopying, microfilming, and recording, or in any information storage or retrieval system, without written permission from the publishers.

For permission to photocopy or use material electronically from this work, please access www.copyright.com (http://www.copyright.com/) or contact the Copyright Clearance Center, Inc. (CCC), 222 Rosewood Drive, Danvers, MA 01923, 978-750-8400. CCC is a not-for-profit organization that provides licenses and registration for a variety of users. For organizations that have been granted a photocopy license by the CCC, a separate system of payment has been arranged.

Trademark Notice: Product or corporate names may be trademarks or registered trademarks, and are used only for identification and explanation without intent to infringe.

Library of Congress Cataloging-in-Publication Data

Names: Brahams, Stanley Bernard, author.
Title: The fundamentals of quality assurance in the textile industry /
Stanley Bernard Brahams.
Description: Boca Raton, FL : CRC Press, 2017. | Includes bibliographical
references and index.
Identifiers: LCCN 2016023859| ISBN 9781498777889 (hardback : alk. paper) |
ISBN 9781315402505 (ebook)
Subjects: LCSH: Clothing trade--Quality control. | Textile fabrics--Quality
control.
Classification: LCC TT497 .B685 2017 | DDC 687.068/8--dc23
LC record available at https://lccn.loc.gov/2016023859

Visit the Taylor & Francis Web site at
http://www.taylorandfrancis.com

and the CRC Press Web site at
http://www.crcpress.com

To my grandsons Oliver and Rory,

I hope this book will be an inspiration in your lifelong journey in the pursuit of knowledge that will benefit you and be of benefit to others.

Contents

Preface xi
Acknowledgments xv
Author xvii
Introduction xix

Chapter 1 Who Is Responsible for Quality? 1

    Defining the Role of the Quality Controller 1
    Company Profiles 2
        International High Fashion Brand with Shops and
        In-Store Concessions - Company A 2
        Chain of Departmental Stores, Catalog Companies,
        and Internet Shops - Company B 2
        Import Agent - Company C 3
        New Label - Company D 3
    Job Brief 4
        Candidate Profile 5

Chapter 2 Risk Analysis 7

Chapter 3 Three Fundamentals for Effective Control of the
Supply Chain 11

    Supply Chain 11

Chapter 4 Writing Procedures for the Supply Chain 13

Chapter 5 Specifications 17

    Specification Examples 21
        Reefer Coat 22
        Men's Short-Sleeve Knitted Shirt 36
        Ladies' Fashion Trouser 45
        Rucksack 54

	Men's Cardigan 63
	Girls' Dress with Full Skirt 68
	Ladies' Unlined Motorbike Jacket 75
	Ladies' Tie Dress with Bodice Seams 82
	Sending Specifications 88
Chapter 6	Fit and Fit Sessions 89
Chapter 7	Fabric Specification and Performance 93
Chapter 8	Fabric Testing 97
Chapter 9	Supplier's Manual 101
	Sampling Procedures 102
	Fabric Minimum Performance Standards 103
	Basic Size Charts and Specifications for Core Lines ... 108
	Manufacturing Guidelines 125
	Packaging and Presentation 125
	Trouser Hanger with Bar 130
	Trouser's Inside 131
	Standard Carton Specification 147
	Acknowledgments 148
Chapter 10	Product Development 149
	Review of a Jeans Range 149
	Girls' Dress Development 155
Chapter 11	Managing the Critical Path 163
Chapter 12	Sample Reports and Approval 169

Chapter 13 Assessing and Working with Factories 175

    Report from China 177
        Factory A 179
        ITR 318 180
        ISK 326 Black Mini with Diamante's 180
        ITR 318 for Bigger Sizes 180
        Factory B 181
        Factory C 181
        Factory D 182
        Factory E 182
        Factory F 182
        Factory G 183
        Factory H 183
        Factory I 183
        Factory J 184
        Summary 185

Chapter 14 Inspection of Merchandise 187

    Acceptable Quality Levels 187
    Examining Garments 190
    Factory History 194

Chapter 15 More Preventative Action 197

    A Case Study - Review of a Company's Quality
    Assurance 197
        Sizing and Grading 197
        Quality Control Directives 198
        Supplier Manuals 198
        Product Data Management 198
        Summary 199
    Inviting Factory Personnel to Your Head Office 202
    Faulty Returns 202

Scary Moments 203
A Trip to the Stores 204
Benchmarking 204
Team Building 205
Odd Bedfellows 206

Summary 209
Index 211

Preface

When I joined a national catalog company as quality assurance manager, the buying director came in to see me and pinned the following notice on the wall: "The bitter taste of poor quality outlasts the sweet taste of cheap prices."

This was a warning about the dynamics that operated in this and most companies. He made it clear that I had the final word on what was acceptable, not the buyers, and I reported directly to him, but he knew pressure would be put on me by the buyers to accept quality that I considered unacceptable or substandard, and if that happened, it would be my responsibility if this resulted in customer complaints. He trusted his buyers to negotiate the best prices, with the right suppliers, but knew that in some instances the temptation of cheap prices could override sound commercial judgment. Theoretically all the factories we deal with should supply us with equally good quality, but in practice we have to deal with suppliers of varying capabilities; the reason usually is that they make certain goods that others can't supply at a competitive price or don't have the right machinery. Once gaining a foothold in your company, these suppliers will try and compete for more of your business, and because their prices are cheaper, buyers will be tempted to give them a larger slice of the cake. The supplier will promise to improve their quality as we give them more orders, but can a leopard change its spots and what should quality controls response be in this type of scenario? First you do not want to appear to be standing in the way of the company buying goods more cheaply and improving profit margins, but equally using suspect suppliers can be the beginning of the slippery slope to poor unacceptable quality, and the role of quality assurance is vital to act as a counterbalance when these situations arise, devising a strategy to ensure that these suppliers comply with your company's standards. However technically proficient and knowledgeable you are in patterns and garment construction, the role of quality custodian is the most challenging and important part of a quality controllers job as you may at times appear to be a lone voice in saying no when you believe the business will be harmed.

It is worth taking a moment to reflect on the recent history of the clothing industry when approximately 80% of garments sold in either Europe or America were made in those countries and sales, design, and production were usually managed under one roof. Owing to the proximity of design and production, quality problems were quickly identified and resolved. Today production can be thousands of miles away, in a different time zone, in a country whose first language and culture are different from yours; I don't think you could make things more difficult if you tried.

Stanley Bernard Brahams
Member of the Apparel Technical Design Association
Los Angeles

Acknowledgments

To the publishing team at Taylor & Francis Group/CRC Press for giving me the opportunity to pass on my experiences and hopefully some wisdom to new generations starting a career in the textile industry.

To my wife Andrea and family for their encouragement and support, and a special thank you to my son-in-law James Suryomartono for his cartoons capturing the funny side of stressful moments.

Author

Stanley Bernard Brahams attended Jacob Kramer College Leeds and earned city and guilds of London Institute Cutting and Tailoring and city and guilds of London Institute Technicians part 2 certificate in clothing manufacture. His early career was spent in clothing manufacturing, working mainly in the design sector. After completing the city and guilds course in men's tailoring, he progressed to working as a pattern cutter/designer specializing in clothing technology and pattern construction. As manufacturing declined in the United Kingdom, he moved to the retail sector working for a variety of companies in the field of garment technology and quality assurance. He worked closely with buyers and suppliers to improve the quality of the products, establishing brand loyalty with their customers. As products were increasingly sourced offshore, his role became pivotal in ensuring that the supply chain worked effectively. His responsibilities included creating detailed specifications covering all aspects of the products and approval of final samples before the start of production. This was done within a timescale to ensure that the shipment was delivered to the customer on the agreed date. He visited and worked with factories in most of the major offshore textile producing areas.

Introduction

The dictionary definition of quality is excellence, superiority, and class, but this can be misleading when applied to the subject of quality assurance. In business, quality assurance refers to defining the level of the product quality and ensuring that all deliveries meet that standard. The quality of the product is decided by your target customer and the selling price. The role of quality assurance is to ensure that once a specification has been agreed, every product and every production run meets that standard. If you sell cheap affordable clothing, you still aim to be the best in that market and you should always look to improve your quality where possible.

Customers perception of quality can be very different, and comments you are very likely to hear are, "I always buy my clothes from Marks & Spencer because their size 12 always fits my figure perfectly." "I buy all our family clothes from Wal-Mart because when tumble dried they hardly need ironing." Many people favor a brand that celebrities are seen wearing, and this would be their definition of quality.

I have worked for several various types of companies as a garment technologist and quality assurance manager and gained valuable experience over the years in how to put good quality assurance procedures into practice. Quality control is the day-to-day implementation of quality assurance, and to succeed you have to be very proactive at all levels. Quality controllers need to learn as much as they can about all aspects of the business as they will need to liaise with all departments in the company as well as customers and suppliers.

The subject of quality assurance is not often taught in fashion colleges, possibly because it is not recognized as being one of the creative disciplines; however, quality assurance does help shape and define a company's image and future success.

Many companies prefer to train their quality controllers in house, and when joining a company you adopt their methods in the beginning. As you become familiar with the job and the company, you may then start to question how you can improve your role in the business and then you might start to question if the company procedures are correct. Companies may be very strong in certain areas, but very weak in others.

This book introduces students, graduates, and companies that are expanding their production offshore to the concept of how quality assurance has to become an integral part of the business. It encompasses procedures that have been adopted internationally by the clothing industry. The aim of this book is to cover the following:

- The tools and the best way to use them
- How to avoid pitfalls
- Dos and don'ts
- Best practices
- Establishing and building relationships with suppliers

Quality assurance can be described in simple terms as common sense. You want to buy a product from a factory, tell them what you want, see a sample before they start production, and if you're happy with the sample tell the factory to make the bulk and deliver to your warehouse by a certain date. When you receive the delivery, open a few cartons to make sure the quality is correct and then send the goods out to your shops. In practice it is not that simple for the following reasons:

- How do you tell the factory what you want, in what format, and how much information do you need to send them?
- How do you set a timetable for receiving the approval samples and production and track their progress?
- How many samples should you request from the factory?
- What format should the sample report take?
- How do you ensure that the fabric and trims are correct and what testing should you do?
- When you inspect a delivery how do you get an accurate assessment of the overall quality of the shipment?
- How do you maintain consistency of quality at all times?

Investment in quality control is a form of insurance, but it does not wait for disaster to happen then compensates you; it works to prevent the disaster from happening in the first place. If implemented properly, it is continuously maintaining, servicing, and repairing your supply chain.

The goods you buy and sell are the lifeblood of your company, and you need to do everything possible to protect the supply chain and to ensure that the level of quality stays consistent.

The aim of this book is to show the nuts and bolts required for effective quality control and how written procedures are the key to good quality assurance, but most importantly, it is how we implement them day by day. This is a hands-on book helping graduates and companies through each procedure, explaining the philosophy behind it, and how all parts link together creating effective quality assurance and showing how it helps to make a business successful.

1

Who Is Responsible for Quality?

DEFINING THE ROLE OF THE QUALITY CONTROLLER

The process of range building, sourcing, planning, and controlling production varies from company to company, and the responsibility for quality assurance also depends very much on how the company operates. It is fair to say that quality is everyone's responsibility; however, individuals with specific responsibilities have to be appointed. There are several titles used in industry to describe this role:

- Quality assurance manager
- Quality controller
- Technical manager
- Product development manager
- Technical designer
- Garment technologist

The role can be divided into two main areas of responsibility: technical and quality auditing.

1. Technical responsibilities include patterns, size charts, specifications, fabric approval, organizing fittings, and keeping up to date with new technology.
2. Quality audit's responsibilities include assessment of factories, dealing with quality issues, analyzing why faults are happening, and checking that the production is correct.

In larger companies, these two areas of responsibility are sometimes separated, but in smaller companies, one individual or department has the responsibility for both. Whichever type of company you work for, this role is a key position; responsibilities vary from company to company, but the objectives are exactly the same. In this book, I'm combining the two roles, so that you have an overview of the total picture of all factors involved in quality and that you should have the ability to influence internally and externally to meet business objectives. Most importantly, you learn to take ownership of the quality standards and issues of the suppliers with whom you are working.

COMPANY PROFILES

When looking for their first job, many design graduates will be happy to take any employment offered to them, as long as it's in the fashion business, but I think that it pays to be focused on the branch of the industry that I would really prefer to work in. Below, I have listed four different types of companies and explained how the role of the technologist may vary within each of them.

International High Fashion Brand with Shops and In-Store Concessions - Company A

This type of company will include head office, employing stylists, designers, pattern cutters, with its own sample room. This company will most likely supply patterns to the factories and employ its own fabric buyers and technicians. Most production will be made offshore in many different countries, and it may have offices based locally in those countries to source and control production. It will also include head of quality and technical services, possibly at the director level, with a team of technologists, each responsible for a garment category, for example, woven top, trousers, knitwear, dresses, tailoring, nightwear, and underwear. Technologists will liaise with their head office fabric department, working closely with and visiting factories in the many different countries. Inspection of finished goods will be done by a quality control (QC) team at the factory or at the company's warehouses.

Chain of Departmental Stores, Catalog Companies, and Internet Shops - Company B

This type of company will sell a diverse product range for men, women, and children of all ages.

The company may employ designers, but usually, the buyers will put their own ranges together by visiting fashion and trade fairs and by working with their suppliers, researching new trends.

Each buying department may have a technologist as part of its team, who will report either to a QC manager or to the buying manager. Fabric may be tested in-house or by an accredited lab, and the technologist may have the responsibility to ensure that the testing is carried out satisfactorily. Delivery inspections will be done by a team at the company's warehouses or by the technologist at the factory.

Import Agent - Company C

The main business of import agents is to provide supplies to chain of departmental stores, catalog companies, and internet shops. They will be working with factories in many countries, and today, they employ their own designers, who, as part of the service, will put new ranges together or source styles that the customers want as part of their range. They will offer a full service of design, sourcing fabrics, sampling, production, shipping, and warehousing and delivery inspection.

The technologist will usually have the responsibility for the complete package of all products for individual customers, liaising with both their QC and their buying departments. They will ensure that fabrics and trims are tested and approved to the customer's standard and will check approval samples before submitting to the customer and attending fit sessions with the customers' QC. By working closely with the factories, they will visit them on a regular basis, finalizing approval samples and checking production, sometimes with the customers' own QC personnel.

New Label - Company D

The last example is a company that has been trading only for several years or even possibly for just a few seasons. They are design led and successful owing to their designs being original, commercial, and correctly priced, and they were fortunate in teaming up with the right manufacturers. They find themselves in the enviable position of an increasing demand for their designs. Success may come quickly, but once it is achieved, maintaining it is the difficult part. A new company that grows rapidly and whose priority has been to design and market will not have had time to think too much about organization; however, as they continue growing, their product range will increase and so will the quantities they sell, and they will need to look to source new factories. This is the time they will need to introduce procedures for controlling the sampling and production to ensure that they maintain the quality that has given them their success. Up to this point in time, the designers will have probably traveled to the factories with ideas and sketches to work on the new ranges, developing a close working relationship with the manufacturer. At this stage in the company's growth, a technical designer/quality controller would take responsibility for liaising directly with the factories for creating technical packs, finalizing fit/sealing samples, and controlling production.

JOB BRIEF

Below is an actual job brief for a senior technologist, advertised by a recruitment agency in London. It clearly shows that the successful applicant will be expected to be involved in all aspects of sampling and production, working closely with all departments. Although the technologist will report to the technical manager, the company will expect the technologist to take day-to-day responsibility for all quality issues.

In this book, I endeavor to cover all aspects of this job description, which are as follows:

- Provide full technical management for the supplier.
- Own the specific technical elements of the product technical specification package: creating and managing the construction details, size charts (utilizing when required CAD images and drawing), key technical areas, presentation, wash care, and packaging.
- Attend fit meeting and line reviews with all customers. This includes measuring and preparing all samples (with help from a junior/freelance QA).
- Provide a liaison with vendors on all aspects of fit, construction, and technical and quality-related issues, which will involve some overseas travel to the Far East.
- Create, maintain, and manage the library for the standards of features, construction, trim, components, labels, packaging, and so on.
- Review supply base, with close liaison to monitor supplier performance, and introduce new suppliers, as required.

- Constantly monitor and improve return rates.
- Organize all fit meetings and manage workload within critical path.
- Manage compliance requirements.
- Identify and resolve any quality issues with shipments that may occur.
- Review on-going processes and bring improvement in the quality of the brand.

Candidate Profile

- Computer literate to include Microsoft Office
- An excellent knowledge of garment construction
- Knowledge of current fashion trends
- Organization skills
- Attention to detail
- Ability to manage time and workload
- Flexibility
- Competence with multiple products
- Able to work with all markets for production and customers
- Experience of auditing processes
- Full understanding of lab tests
- Strong communication skills

2

Risk Analysis

Decisions on quality issues are rarely black or white, especially when they involve textiles. First, we work with materials that are often not stable and will stretch, shrink, or distort during cutting, sewing, or pressing; however, carefully, we try to prevent this from happening. Second, no two machinists sew in exactly the same way, and sometimes, a faulty batch can be traced back to one machinist. Making decisions on the acceptability of samples or production is often difficult, and "commercial decisions" have to be made, which must always be in the best interest of the business. Here are two examples of faults on a delivery of garments, which require a decision to accept or reject. Seven belt loops were specified on a trouser, but the production comes with six; four buttons were specified on a neck fastening of a polo shirt, but the production comes with three. Technically, these garments are wrong, but should they be rejected? The technologist's first job is to check that the garments are still fit for purpose—do the six belt loops still support the trouser properly with a belt and is the neck opening with three buttons still large enough for the customers to get over their head? If your decision is that the garments are still fit for purpose, you would inform the buyer about the differences between the agreed specification and the final production and your decision that they are still commercially acceptable. Now, with this scenario, the buyer could decide either way. If he or she needs the stock, then he or she will be happy to accept the deliveries with your assurance that the garments are still fit for purpose and that the customers would never probably realize the difference. However, the buyer could decide to reject the deliveries and make it a quality issue for several reasons. He or she might not need the stock or intend to move production to another supplier; in this instance, it is the buyer's choice to decide.

Let's visualize another situation: In this instance of a sweatshirt style, the sealing sample has been approved and production starts; the factory notices that the zip insertion has caused the front edge to be very wavy, much more than the approved sample, and possibly, a large quantity has already been made. It might be due to the fact that the production fabric has more elasticity than the fabric used for the sealing sample or the tension is wrong on the sewing machine, or it might be an operative error. Now, the question arises: what to do with the ones already made? The last thing the factory wants to do is to stop production or repair the faulty ones, as it cost time and money; therefore, instead of informing the customer, a decision is made by the factory management to let the faults go. You might have an idea as to where this is going; when the factory sends production samples, it will select the best ones. Back at the head office, these samples will be approved and the delivery will be given the okay to ship. When the faulty garments are found in your warehouse or on the shelves in the store, the factory's comment will be, you approved the production samples! This can then lead to a long protracted negotiation about who is responsible for accepting the poor quality and the expense of withdrawing stock for a 100% inspection.

Of course, if you happened to be there in the factory at that time, you would insist that production is stopped and find the cause of the fault and set a standard for what was acceptable.

However, the reality is that if we don't have our own people in the factory for the majority of the time, we might not know of a problem for days or weeks. The challenge is to work with factories to ensure that they see quality through your eyes and make the decisions you would make, even if you are thousands of miles away, and if they are unsure of the right course of action, they should contact you immediately.

Usually, but not always, new season's ranges will include styles that are carried forward from previous seasons, because they have been extremely good sellers with low returns. If you are using the same factories, there should be no obvious problems. The buying department will also source from new factories that should be made aware of the procedures during the buyer's visit and of the critical path that your company follows (see Chapter 11). If these factories have no previous history with your company, then there will be a question mark, concerning their ability and willingness to work according to your methods. If possible, an assessment should be done by the quality control department before orders are placed.

If this is not possible, a small trial order should be placed before you commit yourself to placing a larger quantity.

New suppliers as well as companies are continuously looking at new fabrics and styling, which might require testing for laundering and performance properties. Selection samples are usually made by the factories after the buying teams visit, and this might be your first opportunity to check their quality and workmanship. You would assume that if they are selection samples, they should be well made, but this is not always the case, and if you find bad sewing faults at this stage, alarm bells should ring. For example, if there are ranges of jeans or trousers from a new factory, check the construction and particularly the front zips. Pull the zipper up and down and look at the stitching. Zips, especially on jeans and trousers, are a big reason for customer returns, and some factories will use poor-quality zips to reduce the cost, and I have even seen fake YKK zips sewn into jeans (see "Report From China" in Chapter 13). If you have serious doubts about the quality of zips that a supplier is using, they should be independently tested. It is the nature of our business that companies will constantly be sourcing new factories, materials, and design features, and because of this, it is necessary to invest the time at the beginning of the process in suppliers to avoid problems later in the supply chain. Quality assurance should always sit in on selection meetings with the buyers and merchandise team; it is very useful to be on hand if there are any queries about returns and quality issues.

These procedures should be an established practice in your company and, in fact, the majority of buyers would welcome your involvement at this stage, as their priority has been price and range building, and you are their first line of defense against poor quality.

3

Three Fundamentals for Effective Control of the Supply Chain

- **Procedures**
- **Specifications**
- **Critical path (timetable)**

These three fundamentals work together and are interconnected; if one is missing or not working effectively, the other two become ineffective.

Outsourcing companies as well as contract manufacturers have to be aware of everything that is going on and efficiently control what is happening. With the good use of planning systems, you can see where a problem might appear before it actually happens.

SUPPLY CHAIN

The supply chain is the system of organization involving the people, information, and resources to move a product(s) from its original concept to the final destination: the customer. The critical path is the timetable needed to achieve this.

The supply chain is essentially a partnership between the buyer, manufacturer, and any third parties involved in various stages of production, and it maps out the responsibilities of all concerned.

Below is an example of procedures required by an importer sent to their factories; its aim is to show how they both need to respond when inquiries are received from customers.

Our company supplies all types of products, and we have built a reputation for quick response on price, quality, and delivery times.

To continue the growth in business, I have drawn up a series of steps that our company and its manufactures must follow to work in partnership to create a successful supply chain.

1. Inquiry from customers.
2. Within 24 hours, our company will send to its suppliers (by an e-mail) a package containing the following:
   a. Style details
   b. Size chart
   c. Fabric details
   d. Photos of the garment (if possible)
   e. Indication of quantity (if available)
   f. Request best price and delivery date
   g. *We are competing with other suppliers for the business, so manufacturer should reply, if possible, within 24 hours with price and delivery date.*
3. Customer's sample and/or fabric will be sent as soon as available.
4. Customer will accept price and delivery dates subject to a sample being approved.
5. Manufacturer must then send counter samples (2 samples of size 12, unless otherwise stated, plus the sample report, with all sections filled in: supplier's name, size, style reference, fabric composition, weight, and construction. The samples must be as close as possible to the original/specification.

    If possible, in correct fabric and color or in correct fabric and nearest color. Nearest possible fabric. (Correct fabric and color must be approved before start of production.) Original sample must be returned with counter samples. Our company should receive samples within 8 days maximum from the request being made. This includes 4 or 5 days for making and 3 days for courier service.
6. Customer approves samples. Size breakdown and quantities will be confirmed. Final approved size charts, label, and packaging instructions will be issued. If all aspects of samples are correct, the customer will request production samples. If the sample is approved subject to amendments, preproduction samples will be requested. Three samples should be made: one for the customer, one for the head office, and one to be returned to the manufacturer as your standard for production. *If samples are not approved, new samples will be requested and orders could be delayed or cancelled.*

4

Writing Procedures for the Supply Chain

Any part of a business needs guidelines to operate efficiently, and this is especially true with quality assurance. It is the quality controller's responsibility to ensure that everyone understands and follows those procedures. The aim of quality assurance is to introduce procedures that act as a safeguard, recognizing potential problems before they happen.

Procedures are the route for the critical path, which is the journey from design concept to goods going into the stores. This is a continuous process through the life of the business; the path has many obstacles and dangers, and the procedures minimize the risk, guiding us safely through to the final stages. This is made all the more difficult as we often work against the clock to meet tight deadlines. In such cases, things do go wrong, but we have to deal with problems quickly and get back on track. Procedures are a plan of action, so that when one stage is completed, we automatically go to the next stage, and everyone concerned know what the next stage is and, most importantly, know what is expected of them.

Simply put, quality assurance is the process of specifying clearly what we want and then introducing a series of checks to ensure that's what we are going to get. This is the way how you protect your brand and your company's reputation.

The flow chart in Figure 4.1 is a typical set of procedures to control the sampling and production process. Each stage is a stepping stone on the critical path toward the development and finalization of the product. It is a logical progression to minimize potential problems. Missing any of these stages can create problems with the final production; use this as a guide, but certainly add other procedures that might be considered important to your business. When procedures are established, it will be a reference point

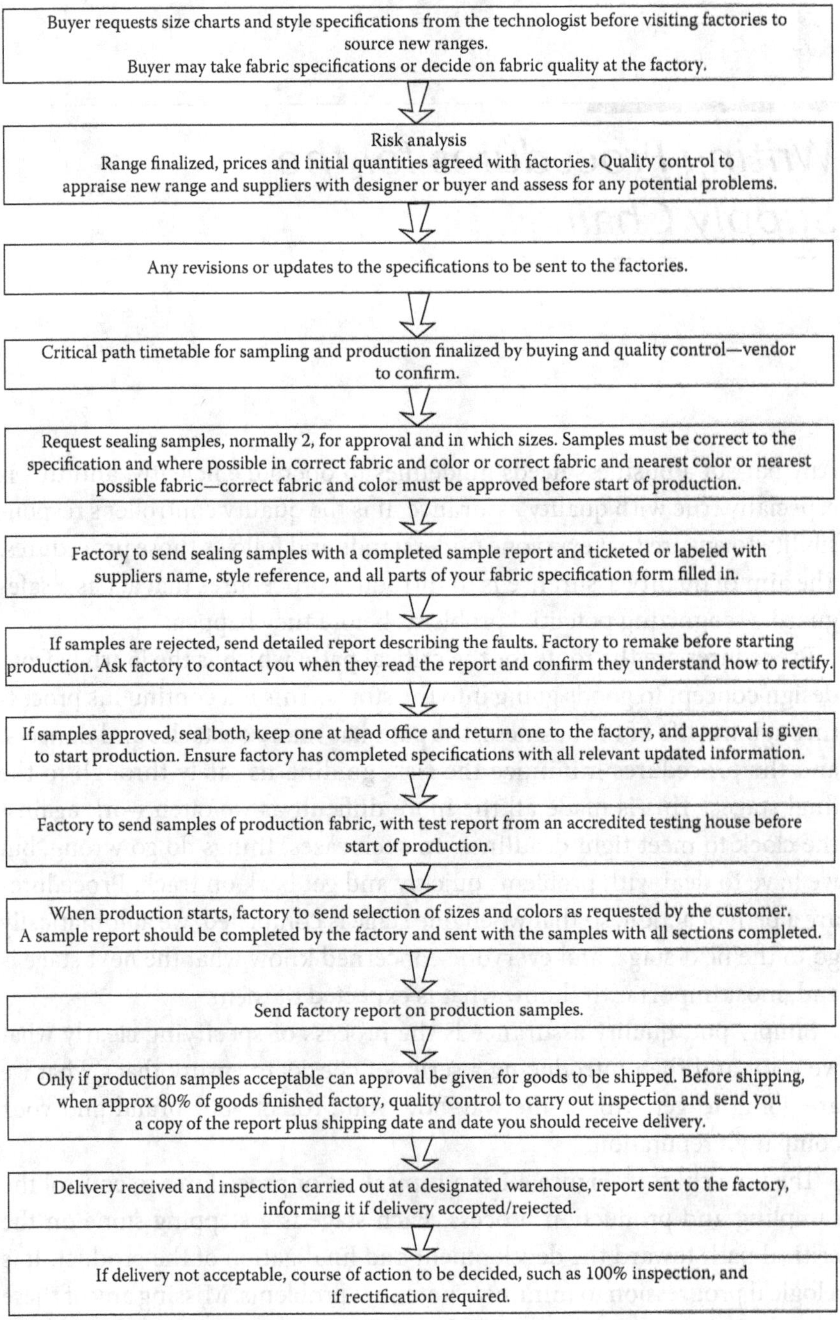

```
┌─────────────────────────────────────────────────────────────────────────┐
│ Buyer requests size charts and style specifications from the technologist│
│ before visiting factories to source new ranges.                          │
│ Buyer may take fabric specifications or decide on fabric quality at the  │
│ factory.                                                                 │
└─────────────────────────────────────────────────────────────────────────┘
                                    ⇩
┌─────────────────────────────────────────────────────────────────────────┐
│ Risk analysis                                                            │
│ Range finalized, prices and initial quantities agreed with factories.    │
│ Quality control to appraise new range and suppliers with designer or     │
│ buyer and assess for any potential problems.                             │
└─────────────────────────────────────────────────────────────────────────┘
                                    ⇩
┌─────────────────────────────────────────────────────────────────────────┐
│ Any revisions or updates to the specifications to be sent to the         │
│ factories.                                                               │
└─────────────────────────────────────────────────────────────────────────┘
                                    ⇩
┌─────────────────────────────────────────────────────────────────────────┐
│ Critical path timetable for sampling and production finalized by buying  │
│ and quality control—vendor to confirm.                                   │
└─────────────────────────────────────────────────────────────────────────┘
                                    ⇩
┌─────────────────────────────────────────────────────────────────────────┐
│ Request sealing samples, normally 2, for approval and in which sizes.    │
│ Samples must be correct to the specification and where possible in       │
│ correct fabric and color or correct fabric and nearest color or nearest  │
│ possible fabric—correct fabric and color must be approved before start   │
│ of production.                                                           │
└─────────────────────────────────────────────────────────────────────────┘
                                    ⇩
┌─────────────────────────────────────────────────────────────────────────┐
│ Factory to send sealing samples with a completed sample report and       │
│ ticketed or labeled with suppliers name, style reference, and all parts  │
│ of your fabric specification form filled in.                             │
└─────────────────────────────────────────────────────────────────────────┘
                                    ⇩
┌─────────────────────────────────────────────────────────────────────────┐
│ If samples are rejected, send detailed report describing the faults.     │
│ Factory to remake before starting production. Ask factory to contact you │
│ when they read the report and confirm they understand how to rectify.    │
└─────────────────────────────────────────────────────────────────────────┘
                                    ⇩
┌─────────────────────────────────────────────────────────────────────────┐
│ If samples approved, seal both, keep one at head office and return one   │
│ to the factory, and approval is given to start production. Ensure factory│
│ has completed specifications with all relevant updated information.      │
└─────────────────────────────────────────────────────────────────────────┘
                                    ⇩
┌─────────────────────────────────────────────────────────────────────────┐
│ Factory to send samples of production fabric, with test report from an   │
│ accredited testing house before start of production.                     │
└─────────────────────────────────────────────────────────────────────────┘
                                    ⇩
┌─────────────────────────────────────────────────────────────────────────┐
│ When production starts, factory to send selection of sizes and colors as │
│ requested by the customer. A sample report should be completed by the    │
│ factory and sent with the samples with all sections completed.           │
└─────────────────────────────────────────────────────────────────────────┘
                                    ⇩
┌─────────────────────────────────────────────────────────────────────────┐
│ Send factory report on production samples.                               │
└─────────────────────────────────────────────────────────────────────────┘
                                    ⇩
┌─────────────────────────────────────────────────────────────────────────┐
│ Only if production samples acceptable can approval be given for goods to │
│ be shipped. Before shipping, when approx 80% of goods finished factory,  │
│ quality control to carry out inspection and send you a copy of the report│
│ plus shipping date and date you should receive delivery.                 │
└─────────────────────────────────────────────────────────────────────────┘
                                    ⇩
┌─────────────────────────────────────────────────────────────────────────┐
│ Delivery received and inspection carried out at a designated warehouse,  │
│ report sent to the factory, informing it if delivery accepted/rejected.  │
└─────────────────────────────────────────────────────────────────────────┘
                                    ⇩
┌─────────────────────────────────────────────────────────────────────────┐
│ If delivery not acceptable, course of action to be decided, such as 100% │
│ inspection, and if rectification required.                               │
└─────────────────────────────────────────────────────────────────────────┘
```

FIGURE 4.1
Procedures to control the sampling and production process.

for all involved with the merchandise. Especially, buying and merchandising departments will want to know the progress of the goods in which they are investing so much money and hopefully make the company a great deal of profit; mistakes are very costly.

Only on repeat deliveries and where the factory has made a style before can some of these stages be missed out, and the decision to do this should only be based on keeping a recorded history of how a factory has performed over a given time.

As a golden rule, approval samples that have major faults, for example, those that measure too big or too small, those that have poor workmanship or fabric, and those that are not of the right quality, should not be accepted, even with the pressure to keep to the agreed timetable. If the factory cannot get the sample right at this stage, then it is very unlikely that it will be correct for production. You should not accept factory reassurances that the production will be correct, but you should insist on seeing correct samples. You need to convince the factory that it is in their interest to produce a sample that is correct before they start the bulk and that it is their responsibility to make the time up later and get back to the agreed schedule.

5

Specifications

Today manufacturing is truly international. Retailers place orders in many countries, and this needs to be reflected in the way we design and communicate product specifications, relying as much as possible on illustration rather than text, because much can be lost in translation. IKEA flat pack assembly instructions are an excellent example, retailing worldwide; they rely totally on illustration, and whichever country they are sold in, the customer can follow the instructions.

The specification can be a complex document with many interrelating parts, and sometimes, at the start of building a range, not all the information will be available. The specification is first initiated when the buyer allocates a reference number, confirms the supplier style, colors, sizes, and fabric details. It is then usually the quality controller's job to be a nuisance and chase (remind) whoever necessary for any missing information to complete the picture.

A specification can originate from several sources; the following are some of the most common:

- Buyer or designer discussing new ranges directly with the factories and arranging for approval samples to be sent to you at the head office. (The buyer may give you samples bought in local stores to create specifications to take with on the buying trip or use the suppliers specification for initial reference.)
- If your company has its own pattern and sample room, the initial specifications will be created at the same time the samples are made.
- Creating new variations from existing style specifications.
- Using existing specifications that require updating.

When you receive initial samples from the supplier for evaluation, please ensure that the buyer has seen and approved each garment first. This will include trying the garments on a figure (see Chapter 6) to ensure that they fit correct and amendments to the size chart and specification might be necessary at this stage.

At this point, I think it is important to mention that, although the quality controller takes ownership for completing the specifications, approving samples and the production, the buyer has the ownership of the actual product. It is buyer's choice of factory, style, and fabric; they have agreed to a price and delivery date; and at the end of the day, they're responsible if the items don't sell. Buyers will expect to be kept up to date on the progress of their merchandise and want to know immediately of any problems you find or changes you might consider necessary. In some instances, an e-mail informing them is sufficient, but if you need to make an important change, go and talk to them first, explain the problem and why you think it necessary to change the specification, because at this point, it's sods or Murphy's law that the supplier will try and increase prices or delay the delivery.

The amount of information that you can include is almost limitless, but time is not! However, you do need a minimum of information, and the priority is to highlight the most important features and characteristics of the product.

I try to include as much detail as possible within the time limit available and this does depend on how quickly and accurately you can draw, and the process will be speeded up if you already have a similar product in your library of specifications. This demonstrates the advantages of using CAD.

I first used CAD when working for a sportswear company, which produced detailed performance garments. I desperately wanted to illustrate the garments and create working drawings, but my hand drawing was very poor. Obtaining a copy of CorelDraw, I began to teach myself to illustrate, practicing as much as possible, evenings and weekends, drawing anything of interest; this included a watch, guitar, mobile phone, and chess pieces. I wanted to learn to draw objects in proportion; if the height, width, and outline were accurate, then all the other features and detail would fall into place.

A good example of this was when I worked for a uniform supplier, which made many types of civil, police, and military uniforms and tunics. Although there are many army regiments, the uniform is basically the same; differences are minor, such as variations in pocket styles and sizes and positioning of braid and badges. Police tunics are based on army tunics, which again have minor differences, so there was often an overlap in styles, and other countries' military uniforms are based on British uniforms. You can now appreciate how new styles could be produced very quickly. As you create new styles, they add to your library of templates. This applies equally to fashion garments, as new styles are usually a reworking or reinvention of past styles.

You may work for a large company that has specialists in different areas of merchandise, and you become familiar with creating specifications for a certain groups of items such as trousers and skirts, ladies tops, or dresses, or may be tailored coats and jackets. It is very possible that because of sick leaves or holidays, you may have to cover for an associate, which helps you broaden your experience of other textiles or you may be in a company where you are the only quality controller responsible for all products. Fashion brands often include bags, hats, and shoes in their ranges, as well as clothing, and it is another string to your bow if you can illustrate these competently.

Today, to encourage customers to spend more money, new styles are fed into the shops on a regular basis. They might have a short shelf life or be very successful. Buyers are constantly traveling to major fashion centers, looking for new ideas and buying garments that they think will fit into their ranges. They will quickly require specifications to send to suppliers that clearly depict the style that is required. Full size chart and packaging and labeling instructions can be added later. The priority is to get a price and then a sample. Suppliers will be chosen from those who can turn samples round quickly, are familiar with the type of garment you are asking them to sample, know how to interpret the specification, and choose a suitable material. The fit and balance of a garment cannot be properly specified; we can only give the main critical measurements of the width and length. Only when we see a garment tried on can we be sure that we are happy with the fit (see Chapter 6). Alternatively, if possible, we can base the new style on an existing pattern block, where we know that customers are happy with the fit. Size chart measurements should be taken from the made-up sample and not the patterns, because those are the dimensions that we want the garment to finish at. Patterns often have built-in manufacturing allowances, and these may vary depending on the fabric's or garment's washing process.

There are issues about size charts that you will find the industry divided on: do we specify the full measurement (total circumference) or half measurement, sometimes referred to as the flat measurement (which is taken as we measure the garment flat on the table). These include bust, waist, hip, hem, bicep, and cuff measurements. Pattern cutters prefer working with full measurements, but factories and quality controllers prefer working with half measurements, as they don't have to double the amount when measuring garments. Whichever method you decide to use, make sure it is clearly mentioned on the size chart.

The other issue is tolerances; now, this can be a tricky one! These are the measurement variations from the size chart that we are prepared to accept. Now, some companies do not indicate any tolerances, but that does not mean that they won't accept any. It just means that they don't want to advertise the fact to their suppliers. Other companies specify plus tolerances but not minus tolerances, their reasoning being that if there has to be variations, they better be slightly larger than smaller. Owing to the instability of fabrics and to the fact that sewing machinists cannot be 100% accurate all the time, manufacturers will want their customers to specify plus or minus tolerances; for example, if a bust grade is 5 cm (2"), they will ask for ±2.5 cm (1"). Can we accept garments that are half a size larger or smaller? The answer is, sometimes, yes, it does depend very much on the style. For larger loose-fitting styles, this tolerance would be acceptable, but for close-fitted styles, no; the manufacturer should be instructed to control the manufacturing better at all stages of production and work to a smaller tolerance. As we are not working with rigid materials, we have to accept some tolerance, but this has to be agreed with the buyer and suppliers and may vary depending on the type of garment.

Specifications are the blueprint of a product, sometimes referred to as the technical pack, and should be as comprehensive as possible. It should be a complete package, covering all aspects of the product, including package and labeling. Core lines or styles derived from core lines need as much detail as possible; they are your basics, the bread and butter lines, and they should be regularly reviewed, adding and amending information, where necessary, helping manufacturers to maintain or improve quality standards and also helping new suppliers to make correct samples with an accurate costing.

Each page should have a heading, to include a date, reference number, style, fabric, factory or customer and buyer, and page number. This will give a unique identity to each specification and each page of the specification. This is especially useful when you amend any part of the specification. The factory will be able to immediately locate the page that needs to be replaced. It could be considered safer to reissue the whole specification with a note of what has been changed and ask the manufacturer to sign a form, acknowledging that it has received the amended specification and that it understands the changes.

SPECIFICATION EXAMPLES

An important aspect of quality assurance is to match products with suitable factories. New factories trying to build up their customer base or factories short of work will sometimes offer to make products that they are not used to making; they're only focused on getting orders, but in the long run, this can waste a lot of time and create many problems. However detailed your specification is, it's the interpretation of the detail that is most important and that can only be done by factories that are constantly making that type of product. For example, a factory making unstructured jackets will struggle making tailored jackets, because it is unlikely to have the right skills or machinery.

It is impossible to work without any form of specification, and a written description or verbal agreement is just not enough. Relying on sending your only sample to a factory 10,000 miles away is also not a good idea. Before a product is finalized, it can be amended many times, and the specification is the only accurate method to note any changes.

The examples that I show here are a combination of complete specifications that include fabric details, chart with all sizes, and packaging instructions. Other specifications are for products in the stages of development where fabric and sizing have not been finalized, and the aim is to quickly get a price and correctly fitting sample. Once these are approved, labeling, packaging, and graded size chart will be added to complete the package. (*Where fabric has not been finalized, the supplier should refer to the supplier manual for minimum fabric performance standards*; see Chapter 9.)

This is another reason for choosing the right factory. When a buyer likes a style but is undecided as to which fabric quality to use, a factory experienced in making similar products will be able to source suitable fabrics and trim. The specification only needs to focus on the style features of the product.

Care and composition labels can be finalized only when the production fabric quality and colors are confirmed. Factories and the buying department will already have a ballpark figure worked out for different methods of packaging and shipping. The buyer can choose between shipping by air and shipping by boat, or a combination of the two.

Reefer Coat

This specification could be based on a similar style garment and the existing specification quickly modified to the new style.

The buyer has specified cotton for the outer, but the weight and construction are to be finalized, and the buyer might decide on a blend of cotton and polyester. Whatever the final fabric choice, we can instruct the factory to refer to the minimum fabric performance for shower-resistant garments and quilted coats copied from the supplier's manual. Rather than trying to put all the details on one page, it is better to create pages for each section of the garment (Figure 5.1).

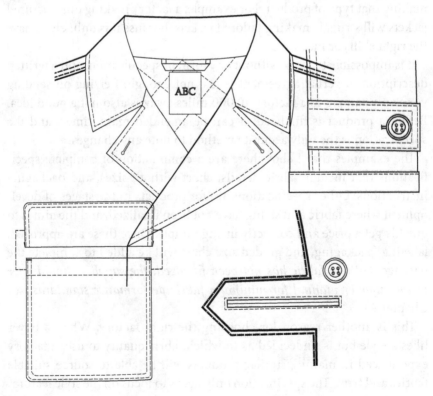

FIGURE 5.1
Reefer coat.

- The first page of the specification, as shown in Figure 5.2, is the minimum textile performance required.
- The second page, as shown in Figure 5.3, is the overall look of the product.
- The third page, as shown in Figure 5.4, shows the garment interlinings.
- The fourth page, as shown in Figure 5.5, shows the front with pocket detail.
- The fifth page, as shown in Figure 5.6, shows the back and cuff detail.
- The sixth page, as shown in Figure 5.7, shows the collar construction.
- The seventh page, as shown in Figure 5.8, shows the design and construction of the inside lining.
- The eighth page, as shown in Figure 5.9, shows individual components of the lining.
- The ninth page, as shown in Figure 5.10, shows details of the buttons and button holes.
- The tenth page, as shown in Figure 5.11, is the garment size chart.
- The eleventh page, as shown in Figure 5.12, shows the size chart, measuring points.
- The twelfth page, as shown in Figure 5.13, shows the garment packaging.

Detail can be copied, pasted, cropped, and enlarged to highlight detail.

SHOWER-RESISTANT GARMENTS ANORAKS QUILTED COATS
Minimum performance requirements

TEST	REQUIREMENT	PAGE 1
Tensile strength BS EN ISO 13934-2: 2014- grab method	150N	
Seam slippage BS EN ISO 13936-1: 2004	6 mm SO 80N	
Seam strength BS EN ISO 13935-2: 1999	SS120N	
Martindale Abrasion BS EN ISO 12947-2: 1999	S/C at 5000 revs grade 3/4	
Martindale pilling BS EN ISO 12945-2: 2000	2000 revs grade 3/4	

DIMENSIONAL STABILITY

Stability to washing BS EN ISO 6330: 2012	+/− 3%
Stability to dry cleaning commercial	+/− 3%

COLOR FASTNESS

Color fastness to washing BS EN ISO C06 2010	Change: 4 Stain: 4 Cross stain: 4/5
Color fastness to dry cleaning BS EN ISO 105-D01: 2010	Change: 4 Stain: 4 Cross stain: 4/5
Color fastness to water BS EN ISO-E01: 2013	Change: 4 Stain: 4 Cross stain: 4/5
Color fastening to rubbing BS EN ISO 105-X12	Dry: 4 Wet 3/4
Color fastness to light BS EN ISO 105-B02: 2014	Std 4 Req 4

FIGURE 5.2
Shower-resistant garments, anoraks, and quilted coats: minimum textile performance required.

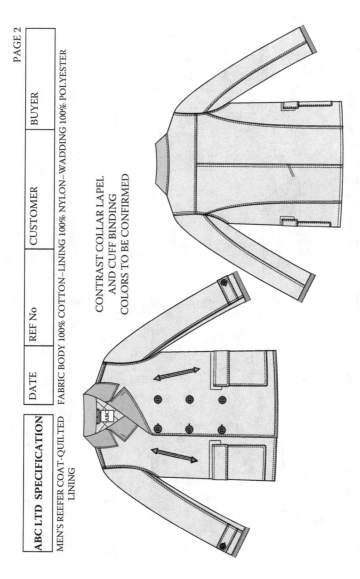

FIGURE 5.3
The overall look of the product.

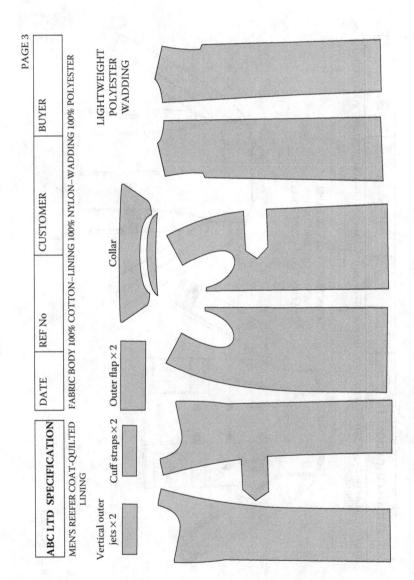

FIGURE 5.4
Garment interlinings.

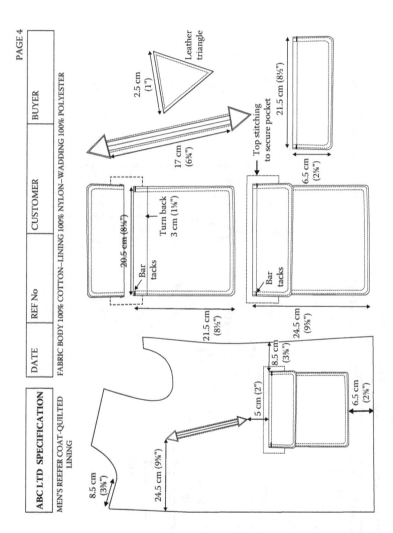

FIGURE 5.5
Front with pocket details.

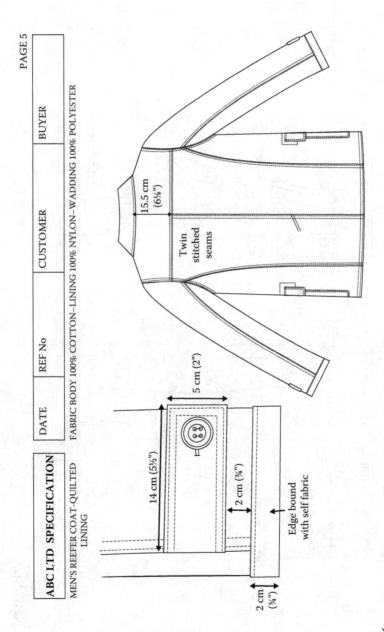

FIGURE 5.6
Back and cuff details.

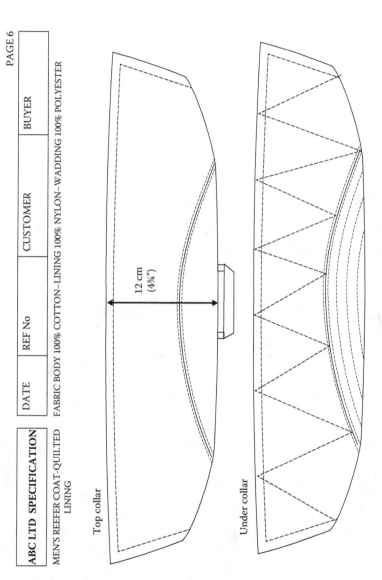

FIGURE 5.7
Collar construction.

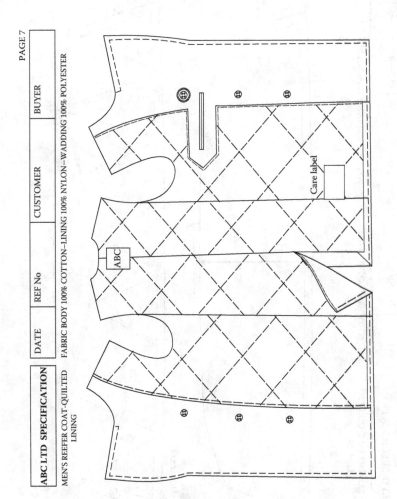

FIGURE 5.8
Design and construction of the inside lining.

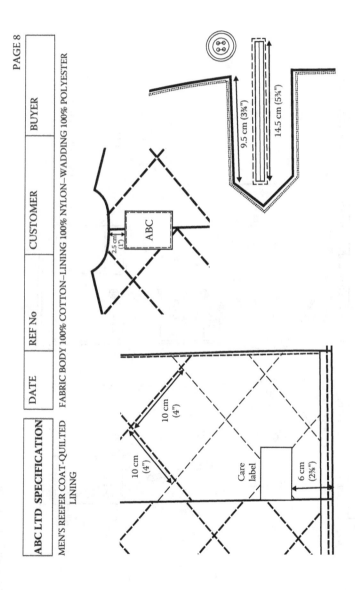

FIGURE 5.9
Individual components of the lining.

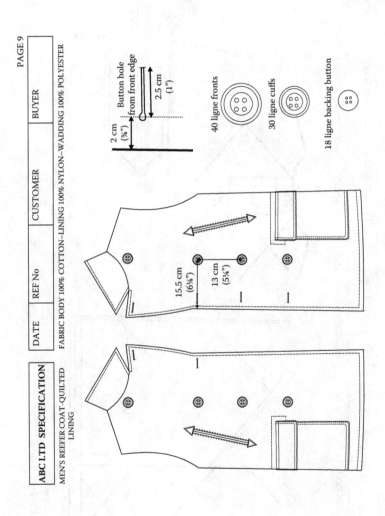

FIGURE 5.10
Details of buttons and button holes.

PAGE 10

Size	S		M		L		XL		XXL	
	cm	inches	cms	inches	cms	inches	cms	inches	cms	inches
A. Chest at base of armhole	53	20¾"	58	22¾"	63	24¾"	68	26¾"	73	28¾"
Coat fastened half measurement										
B. Waist	48	19"	53	21"	58	23"	63	25"	68	27"
Coat fastened half measurement										
C. Hem	51	20"	56	22"	61	24"	66	26"	71	28"
Coat fastened half measurement										
D. Shoulder	16.1	6⅜"	16.4	6½"	16.7	6⅝"	17.3	6¾"	17.3	6⅞"
Sleeve seam to neck seam										
E. Cuff half measurement	14	5½"	15	5⅞"	16	6¼"	17	6⅝"	18	7"
F. Back width	43	17	47	18½"	51	20	55	21½"	58.5	23
G. Full length from neck seam	86	33⅞"	86	33⅞"	86	33⅞"	87	34¼"	87	34¼"
H. Sleeve crown to cuff	67	26⅜"	67	26⅜"	67	26⅜"	68	26¾"	68	26¾"
I. Center vent	26	10¼"	26	10¼"	26	10¼"	26	10¼"	26	10¼"

FIGURE 5.11
Garment size chart.

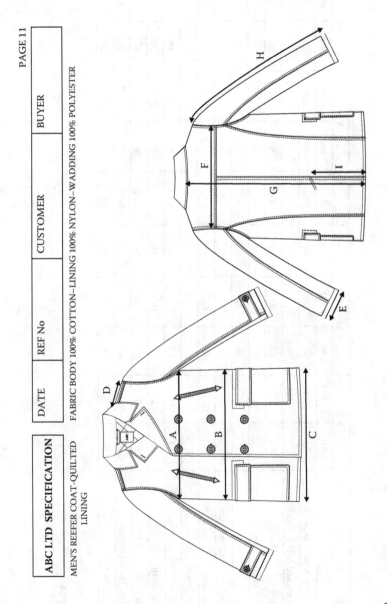

FIGURE 5.12
Size chart measuring points.

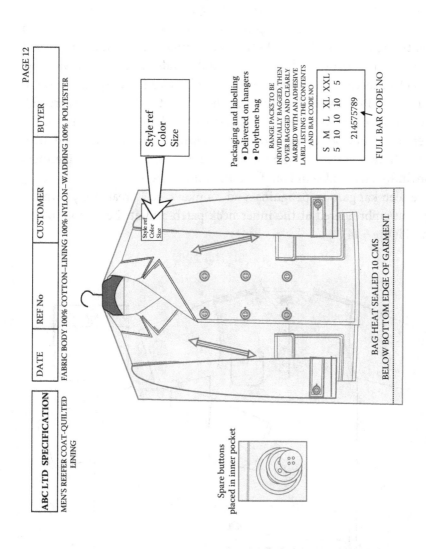

FIGURE 5.13
Garment packaging.

Men's Short-Sleeve Knitted Shirt

This is a popular summer garment and a core line with many companies. The fabric performance, shape, size chart, labeling, and packaging will have already been established, examples of which I have included in the specification. Offered in various designs, the basic features are short sleeves, button placket neck opening, and usually one pocket, made from a light-weight knitted fabric. As this is a summer garment, an important requirement is that the fabric is stable and stands up to regular washing, without distorting, shrinking, or losing color (see fabric specification in Figure 5.15). Prices vary depending on the amount of detail in the garment.

Features have been added to this garment such as contrast tape on the inner collar and side splits and twin stitching at the shoulder and armholes. Many brands dress up their garments with a logo, to make them more individual and stand out from the competition.

I've seen kangaroos, penguins, and monkeys, so I created my own: an elephant embroidered at the inner neck patch and at the bottom of the hem (Figure 5.14).

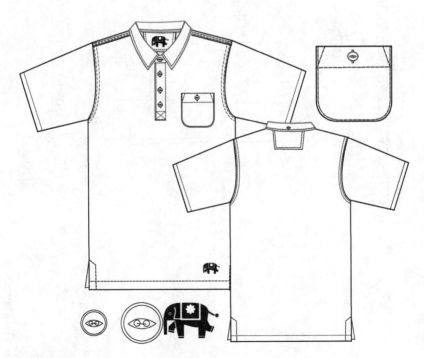

FIGURE 5.14
Men's jersey short-sleeve knitted shirt.

- The first page of the specification, as shown in Figure 5.15, shows the fabric construction and minimum textile performance required.
- The second page, as shown in Figure 5.16, is the overall look of the product.
- The third page, as shown in Figure 5.17, shows the collar and placket details.
- The fourth page, as shown in Figure 5.18, shows the label and embroidery details.
- The fifth page, as shown in Figure 5.19, shows the size chart.
- The sixth page, as shown in Figure 5.20, shows the measuring points.
- The seventh page, as shown in Figure 5.21, shows the packaging instructions.

KNITTED JERSEY TOP
Minimum performance requirements PAGE 1

Fiber composition	65% polyester, 35% cotton
Construction	18 gauge
Threads	1
Weight	150 grams per square meter

TEST	REQUIREMENT
Spirality EN16322-2 procedure C	5% max
Bursting strength BS EN ISO 13938-1 or 2 19999	180 kPa lace 250 kPa
Fabric weight BS EN 12127: 1998	+/−5% of weight as agreed with the buyer
Stretch and recovery BS EN 14704-1:2005 Method A	10%
Martindale pilling BS EN ISO 12945-2: 2000	Knitted 500 revs grade 4

DIMENSIONAL STABILITY

Stability to washing BS EN ISO 6330: 2012	Knitted +/− 5%
Stability to dry cleaning commercial	Knitted +/− 5%

COLOR FASTNESS

Color fastness to washing BS EN ISO 105-C06 2010	Change: 4 Stain: 4 Cross stain: 4/5
Color fastness to dry cleaning BS EN ISO 105-D01: 2010	Change: 4 Stain: 4 Cross stain: 4/5
Color fastness to water BS EN ISO 105-E01: 2013	Change: 4 Stain: 4 Cross stain: 4/5
Color fastening to rubbing BS EN ISO 105-X12	Dry: 4 Wet 3/4
Color fastness to perspiration BS EN ISO 105- E04: 2013	Change: 4 Stain: 4 Cross stain: 4/5

FIGURE 5.15
Knitted jersey top.

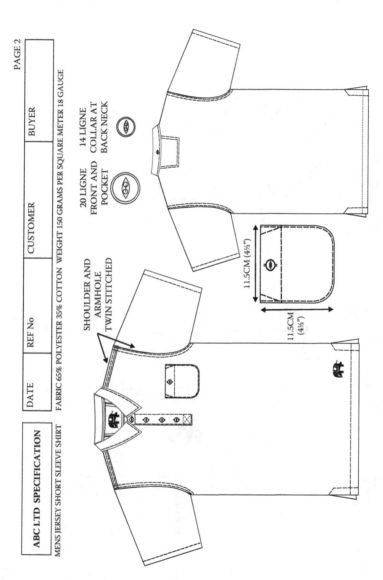

FIGURE 5.16
Overall look of the product.

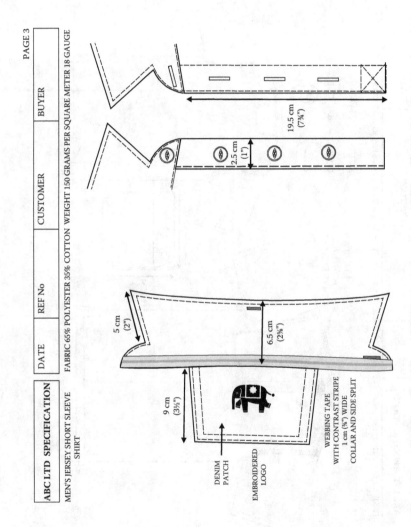

FIGURE 5.17
Collar and placket details.

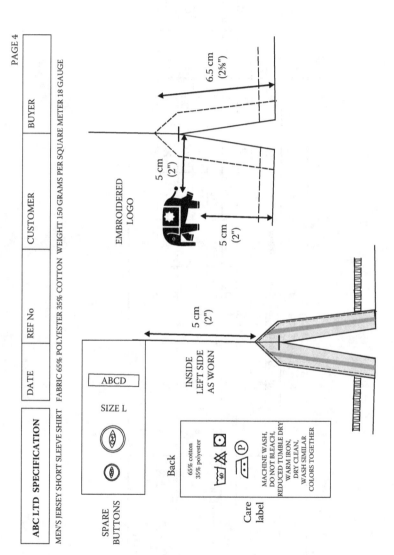

FIGURE 5.18
Label and embroidery details.

PAGE 5

		S		M		L		XL		XXL	
		cm	inches	cm	inches	cm	inches	cm	inches	cm	inches
Chest half measurement	A	48	18⅞	53	20⅞	58	22⅞	63	25	68	27
Hem half measurement	B	48	18⅞	53	20⅞	58	22⅞	63	25	68	27
Shoulder from neck seam	C	13	5⅛	14	5½	15	5⅞	16	6¼	17	6⅝
Sleeve opening half measurement	D	14.5	5¾	15.5	6⅛	16.5	6½	17.5	6⅞	18.5	7¼
Armhole half measurement	E	20.5	8⅛	22	8⅝	23.5	9¼	25	9⅞	26.5	10⅜
Sleeve length	F	23	9	23	9	24	9½	25.5	10	25.5	10"
Full length from neck point	G	76	29⅞	76	29⅞	76	29⅞	78	30¾	78	30¾
Collar center of button to center button hole	H	35	13¾	37.5	14¾	40	15¾	42.5	16¾	45	17¾

FIGURE 5.19
Size chart.

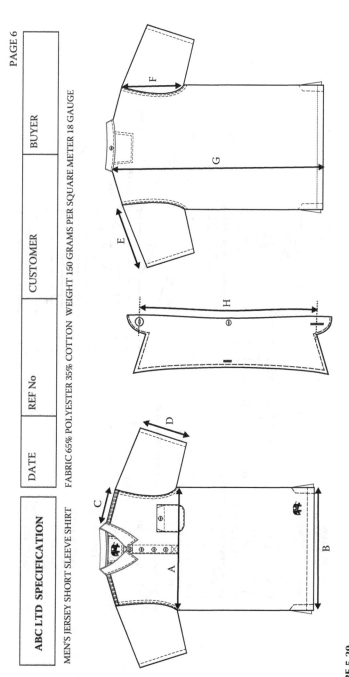

FIGURE 5.20
Measuring points.

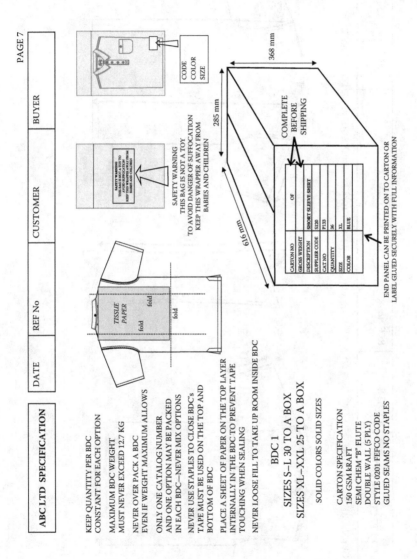

FIGURE 5.21
Packaging instructions.

Ladies' Fashion Trouser

This style could be part of a new seasons range and is offered in several colors. The fabric performance, shape, size chart, labeling, and packaging will have already been established, examples of which I have included in the specification. The style could be tweaked many times, making style changes, but the specification can easily be reworked and quickly changed.

Packaging is sometimes overlooked; factories you have worked with on a regular basis will get to know your packaging requirements, but companies are regularly sourcing new factories, and it is quite surprising how many different ways there are of folding a pair of trousers and placing them in a bag and carton. Even if you are working with a familiar factory, nothing should be left to chance, because if the person who normally looks after your account is away, someone else could take over, who is not familiar with your requirements, and companies have many customers all with different needs. You must instruct your suppliers to follow your specifications, which must be comprehensive and easy to follow, as they are aware of financial penalties or claims against them if the product or packaging is faulty or not as specified, and if a claim is made, they will instantly look for the get-out clause "it wasn't specified" or "the specification wasn't clear" (see section on packaging in "Suppliers Manual" [Chapter 9]; Figure 5.22).

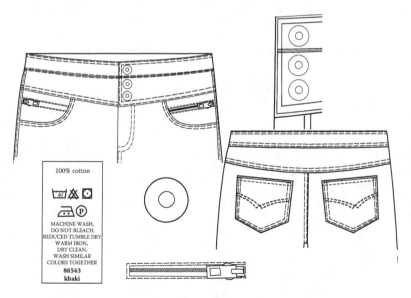

FIGURE 5.22
Ladies fashion trouser.

- The first page of the specification, as shown in Figure 5.23, is the fabric construction and minimum textile performance required.
- The second page, as shown in Figure 5.24, is the overall look of the product.
- The third page, as shown in Figure 5.25, shows the back and front details.
- The fourth page, as shown in Figure 5.26, shows the inner waistband.
- The fifth page, as shown in Figure 5.27, shows the measuring points.
- The sixth page, as shown in Figure 5.28, shows the size chart.
- The seventh page, as shown in Figure 5.29, shows the packaging instructions.

LADIES' TROUSER
Minimum Performance Requirements   PAGE 1

Fiber composition	100% Cotton
Construction	60 × 60 Warp and weft
Yarn count	16s × 16s
Weight	180 grams per square meter

TEST	REQUIREMENT
Tensile Strength BS EN ISO 13934-2: 2014- Grab Method	150N
Seam Slippage BS EN ISO 13936-1: 2004	6 mm SO 80N
Seam Strength BS EN ISO 13935-2: 1999	SS120N
Martindale Abrasion BSENISO12947-2: 1999	S/C at 5000 revs grade 3/4
Martindale pilling BS EN ISO 12945-2: 2000	2000 revs grade 3/4

DIMENSIONAL STABILITY

Stability to washing BS EN ISO 6330: 2012	+/−3%
Stability to dry cleaning Commercial	+/−3%

COLOR FASTNESS

Color fastness to washing BS EN ISO C06 2010	Change: 4 Stain: 4 Cross stain: 4/5
Color fastness to dry cleaning BS EN ISO 105-D01: 2010	Change: 4 Stain: 4 Cross stain: 4/5
Color fastness to water BS EN ISO-E01: 2013	Change: 4 Stain: 4 Cross stain: 4/5
Color fastening to rubbing BS EN ISO 105-X12	Dry: 4 Wet 3/4
Color fastness to light BS EN ISO 105-B02: 2014	Std 4 Req 4

FIGURE 5.23
Fabric construction and minimum textile performance required.

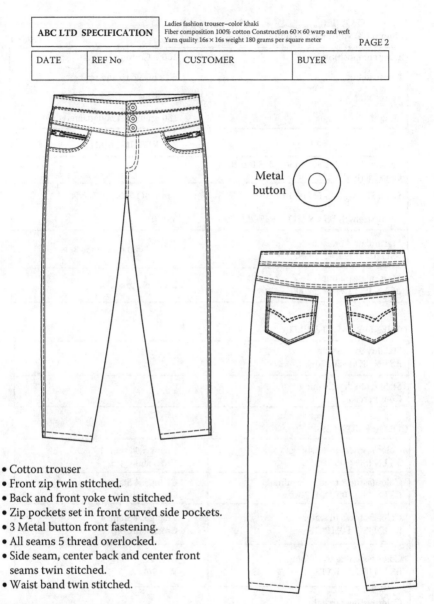

- Cotton trouser
- Front zip twin stitched.
- Back and front yoke twin stitched.
- Zip pockets set in front curved side pockets.
- 3 Metal button front fastening.
- All seams 5 thread overlocked.
- Side seam, center back and center front seams twin stitched.
- Waist band twin stitched.

FIGURE 5.24
Overall look of the product.

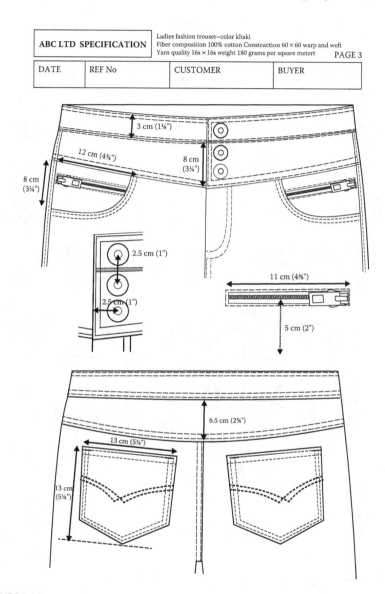

FIGURE 5.25
Back and front details.

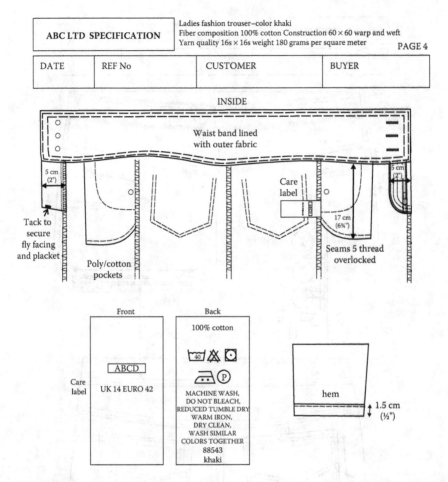

FIGURE 5.26
Inner waistband.

ABC LTD SPECIFICATION	Ladies fashion trouser–color khaki Fiber composition 100% cotton Construction 60 × 60 warp and weft Yarn quality 16s × 16s weight 180 grams per square meter		
DATE	REF No	CUSTOMER	BUYER

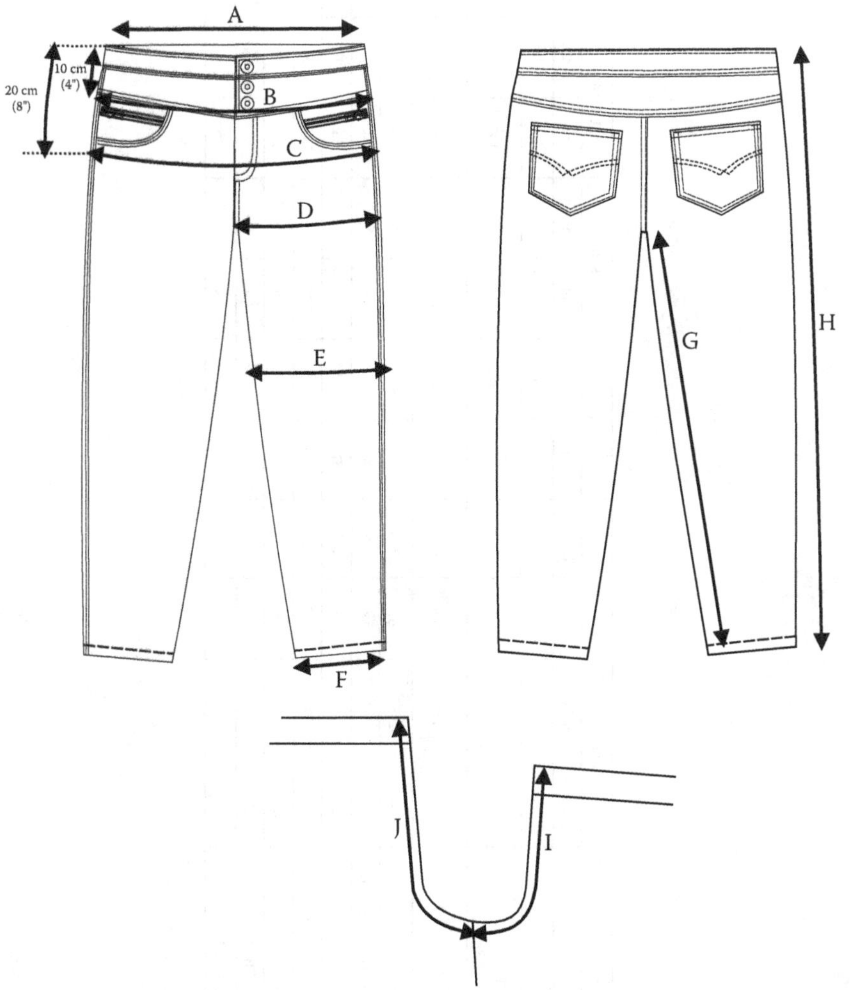

10 cm (4")
20 cm (8")

FIGURE 5.27
Measuring points.

PAGE 6

SIZE	UK 8		UK 10		UK 12		UK 14		UK 16		UK 18	
FULL MEASUREMENTS	cm	inches	cm	inches	cm	inches	cm	inches	cm	inches	cm	inches
A. WAIST OPENING	76	29¾"	81	31¾"	86	33⅞"	91	35¾"	96	37¾"	101	39¾"
B. HIP 10 CM BELOW TOP OF THE WAISTBAND	87	34¼"	92	36¼"	97	38¼"	102	40¼"	107	42¼"	112	44
C. HIP 20 CM BELOW TOP OF THE WAISTBAND	100	39⅜"	105	41⅜"	110	43⅜"	115	45⅜"	120	47⅜"	125	49⅜"
D. THIGH AT CRUTCH	64	25⅜"	67	26⅜"	70	27½"	73	28¾"	76	29⅞"	79	31⅛"
E. KNEE MIDPOINT BETWEEN HEM AND FORK	66	26"	68	26¾"	70	27½"	72	28⅜"	74	29⅛"	76	29⅞"
F. HEM	27	10⅝"	29	11⅜"	31	12¼"	33	13	35	13¾"	37	14½"
G. INSIDE LEG	73.5	29"	73.5	29"	73.5	29"	73.5	29	73.5	29"	73.5	29
H. OUTSIDE LEG, INCLUDING BAND	93.5	36¾"	94	37	94.5	37¼"	95	37½"	96	37¾"	96.5	38
I. FRONT RISE, INCLUDING WAISTBAND	23	9	24	9⅜"	25	9¾"	26	10⅛"	27	10½"	28	11"
J. BACK RISE, INCLUDING WAISTBAND	31	12¼"	32	12⅝"	33	13"	34	13⅜"	35	13⅞"	36	14¼"

FIGURE 5.28
Size chart.

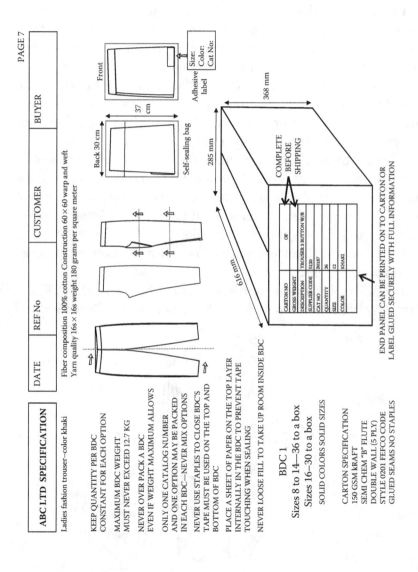

FIGURE 5.29
Packaging instructions.

Rucksack

I did mention earlier that a clothing technologist could be asked to develop a product outside of his or her comfort zone and could be needed to be able to work on and show competence on multiple products.

You never know what is round the corner; my own career took me from men's tailoring to children's wear, from dresses to underwear, from motorbike clothing to nursery products, including pram covers and changing bags and eventually military tunics.

Rucksack is a challenging product but still sewn together like a garment. Rucksacks are designed to be functional, just like parkas and anoraks, which we describe as performance wear. There is a lot of detail on these bags, which must be shown clearly to the maker, especially if we do not have a sample to send, or may be just a photo, which will not show all the details, especially the inside. Time may be against you to show all the details I have shown, but try to include as much as possible. I have not included performance details, but with a little research, I found details of suitable fabrics for this type of product.

Although you try to include as much detail as possible in any specification, the manufacturer might make some minor changes that they believe improves the product. This should never be rejected outright but reviewed, and if considered acceptable, your product details should be amended (Figure 5.30).

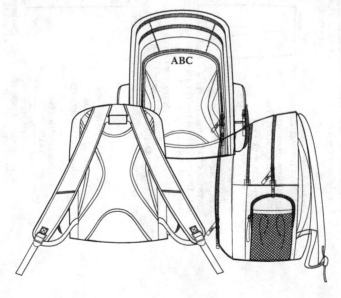

FIGURE 5.30
Rucksack.

- The first page of the specification, as shown in Figure 5.31, shows the overall appearance of the product.
- The second page, as shown in Figure 5.32, shows the measurements of the bag.
- The third page, as shown in Figure 5.33, shows the back and strap details.
- The fourth page, as shown in Figure 5.34, shows the front logo detail.
- The fifth page, as shown in Figure 5.35, shows the side pouch pockets.
- The sixth page, as shown in Figure 5.36, shows the back and middle compartments.
- The seventh page, as shown in Figure 5.37, shows the front compartment.

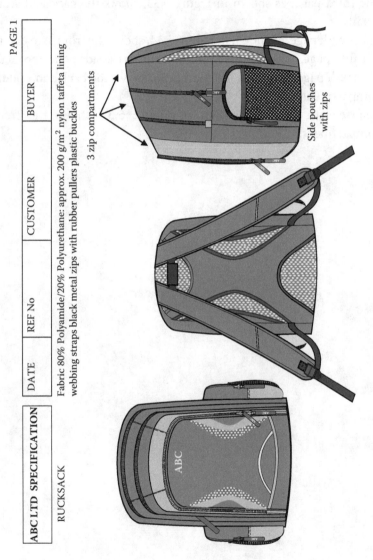

FIGURE 5.31
Overall appearance of the product.

Specifications • 57

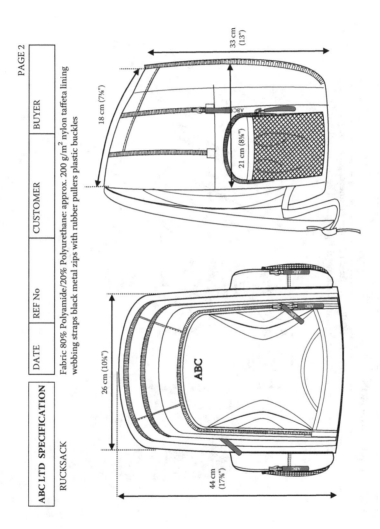

FIGURE 5.32
Measurements of the bag.

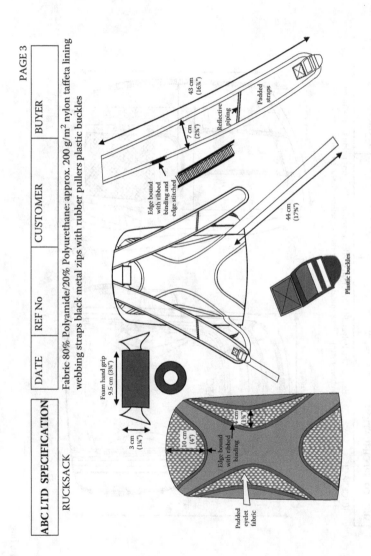

FIGURE 5.33
Back and strap detail.

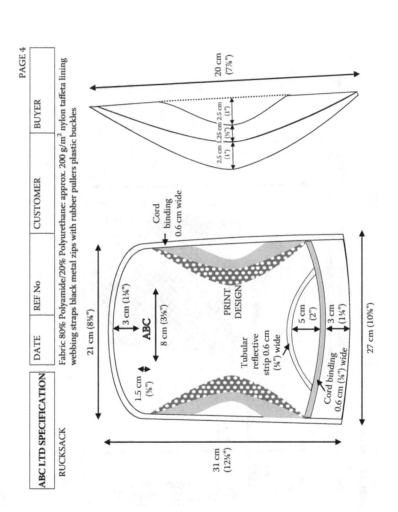

FIGURE 5.34
Front logo detail.

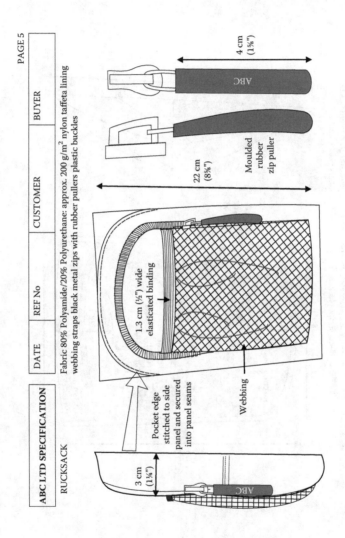

FIGURE 5.35
Side pouch pockets.

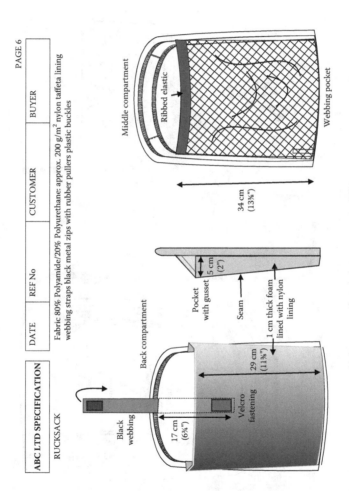

FIGURE 5.36
Back and middle compartments.

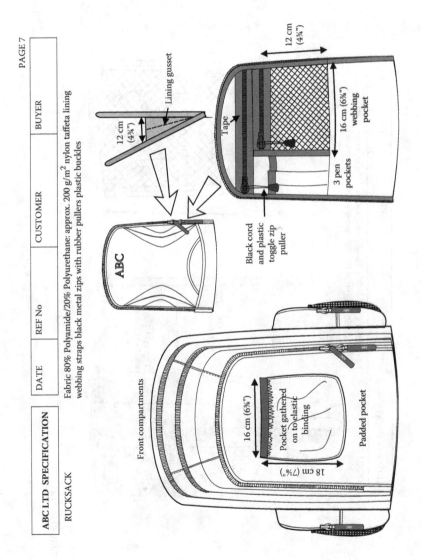

FIGURE 5.37
Front compartment.

Men's Cardigan

A classic men's cardigan is a product that buyers are always looking to source. This specification could be sent out to suppliers as an inquiry for price. The yarn could be wool, acrylic, cotton, or a blend. The buyers will cast their nets as wide as possible. Although they have an open mind about the fiber composition, they don't want to compromise on the stitch construction and weight. This garment has three different rib constructions. The main body, cuffs and bottom welt, and the front band and collar, all are of different sizes. I have specified the number of ribs per 2.5 cm (1″) and the weight per dozen, which tells the manufacturer the yarn thickness, and the photo will also help show the factory the effect that you are looking for.

Whichever fabric is selected, the supplier will be requested to refer to the supplier manual for the minimum performance specification for knitwear, or a copy can be sent with the specification (Figure 5.38).

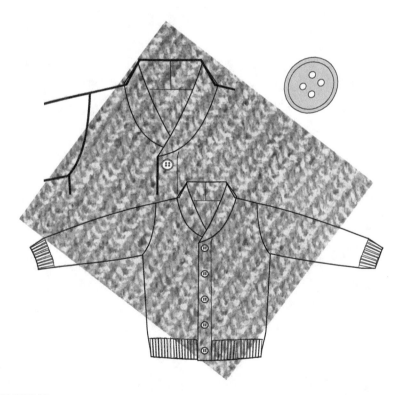

FIGURE 5.38
Men's ribbed cardigan.

- The first page of the specification, as shown in Figure 5.39, shows the style and the details of the ribs.
- The second page, as shown in Figure 5.40, shows the size chart.
- The third page, as shown in Figure 5.41, shows the measuring points.

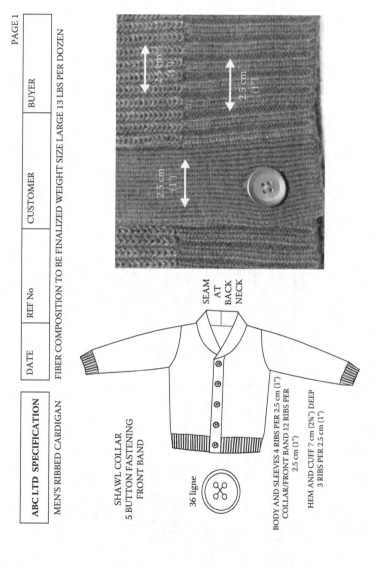

FIGURE 5.39
Style and the details of the ribs.

FLAT MEASUREMENTS	SIZE LARGE	CMS	INCHES			PAGE 2
CHEST	A	62.5	24⅝			
HEM RELAXED	B	55	21⅝			
ARMHOLE	C	29	11⅜			
NECK OPENING	D	27	10⅝			
NECK SEAM TO TOP BUTTON	E	25	10			
FULL LENGTH FROM NECK POINT	F	75	29½			
OVERARM FROM NECK POINT	G	77.5	30½			

FIGURE 5.40
Size chart.

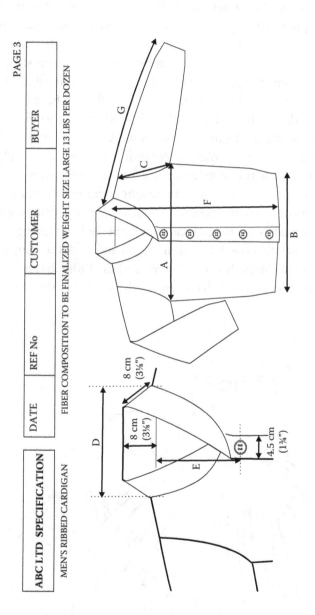

FIGURE 5.41
Measuring points.

Girls' Dress with Full Skirt

Very much a party dress, a buyer might have seen this style in a shop and bought it to try with his or her own children's range. Probably, at first, the buyer will be prepared to place a small order and see how the dress sells and, if sales are good, then include it in the main range. With a target price in mind, the buyer will ask several suppliers for a price and samples. To keep the price competitive, the supplier could be asked to submit suitable fabrics already available at fabric suppliers rather than to have fabric especially made. In these circumstances, the fabric wholesaler would be asked for fabric composition, weight care instructions, and any other information that they have available. Whichever fabric is selected, the supplier will be requested to refer to the supplier manual for the minimum performance specification for wovens, or a copy can be sent with the specification.

There will probably be options of several materials in different weights and composition. It is advisable to have one sample checked by a testing house or, if you have the facilities, check the fabric weight and wash the sample.

The feature of the dress is the full skirt, and the lining has a flounce to emphasize the fullness, so this has been clearly detailed in the specification (Figure 5.42).

FIGURE 5.42
Girls' full skirt dress.

- The first page of the specification, as shown in Figure 5.43, shows the overall appearance and style details of the product.
- The second page, as shown in Figure 5.44, shows the button fastening and braid details.
- The third page, as shown in Figure 5.45, shows the lining construction.
- The fourth page, as shown in Figure 5.46, shows the size chart.
- The fifth page, as shown in Figure 5.47, shows the measuring points.

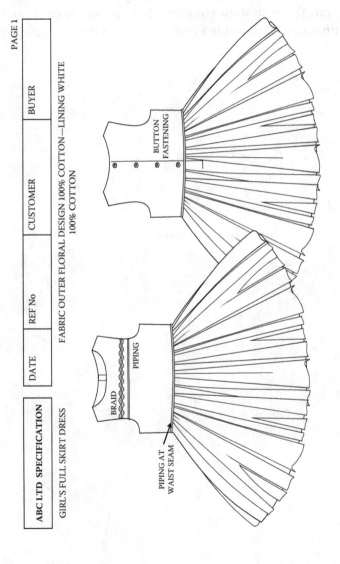

FIGURE 5.43
Overall appearance and style details of the product.

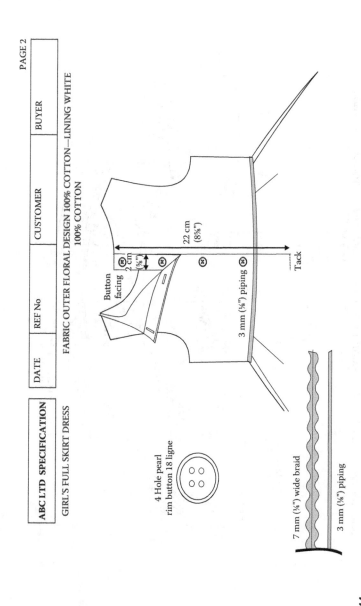

FIGURE 5.44
Button fastening and braid details.

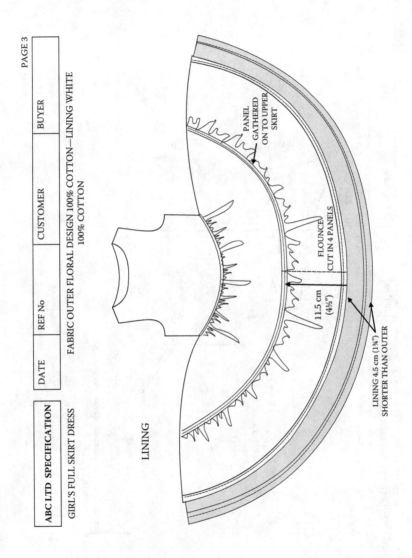

FIGURE 5.45
Lining construction.

PAGE 4

	6–9 MONTHS	CM	INCHES		
BUST AT THE BASE OF THE ARMHOLE ½ measurement	A	23	9⅛		
WAIST ON SEAM ½ measurement	B	23	9⅛		
HEM OUTER FABRIC ½ measurement	C	120	47¼		
HEM LINING ½ measurement		180	70¾		
NECK OPENING	D	12.5	4⅞		
SHOULDER	E	3.5	1⅜		
FRONT NECK DROP	F	2.5	1		
ARMHOLE ½ measurement	G	12.5	4⅞		
FULL LENGTH AT CENTER BACK	H	41	16⅛		
LENGTH NECK TO WAIST SEAM	I	17	6⅝		

FIGURE 5.46
Size chart.

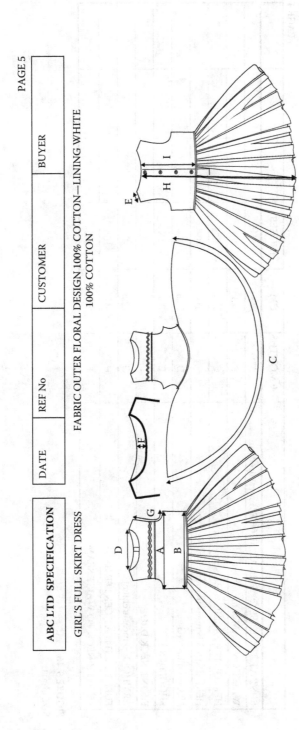

FIGURE 5.47
Measuring points.

Ladies' Unlined Motorbike Jacket

This may have been bought on a buying trip anywhere in the world. The buyers like the style and think it will sell in your stores if they can buy it at the right price and quality. There is only one sample that the buyer wants to keep at the head office. The priority is now to create a specification with fabric composition, style features, and measurements to circulate to suppliers, requesting a price and sample as soon as possible. Once a manufacturer is found and a sample is approved, a size chart can be finalized and a decision made to deliver hanging or flat packed.

With this type of garment, depending on the final fabric and trim choice, you need to check if the garment is washable, dry clean only, or sponge clean. Whichever fabric is selected, the supplier will be requested to refer to the supplier manual for the minimum performance specification for wovens, or a copy can be sent with the specification (Figure 5.48).

FIGURE 5.48
Ladies' motorbike jacket unlined.

- The first page of the specification, as shown in Figure 5.49, shows the overall appearance and style details of the product.
- The second page, as shown in Figure 5.50, shows the collar, facing, and mock leather sleeve and shoulder.
- The third page, as shown in Figure 5.51, shows the pocket cuff and waist band details.
- The fourth page, as shown in Figure 5.52, shows the size chart.
- The fifth page, as shown in Figure 5.53, shows the measuring points.

ABC LTD SPECIFICATION	DATE	REF No	CUSTOMER	BUYER

PAGE 1

MOTORBIKE JACKET
UNLINED

FABRIC BODY 80% COTTON 19% POLYESTER 1% ELASTANE–
MOCK LEATHER SHOULDER AND SLEEVE TRIM 55% VISCOSE 45% POLYURETHANE

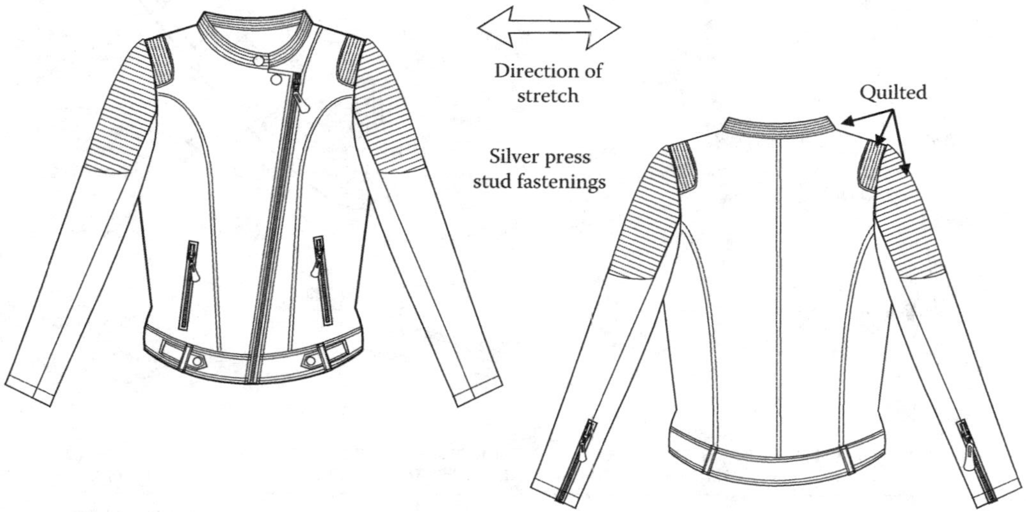

Direction of stretch

Silver press stud fastenings

Quilted

FIGURE 5.49
Overall appearance and style details of the product.

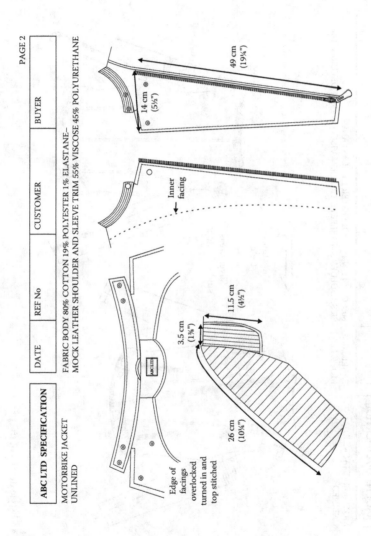

FIGURE 5.50
Collar, facing and mock leather sleeve and shoulder.

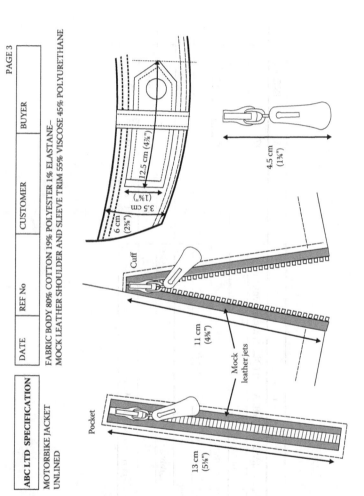

FIGURE 5.51
Pocket cuff and waist band details.

UK 8 USA 12		CM	INCHES	PAGE 4
BUST AT THE BASE OF THE ARMHOLE ½ measurement	A	47	18	
WAIST ½ measurement	B	42.7	16⅝"	
HEM ½ measurement	C	50	19⅝"	
BICEP AT BASE OF ARMHOLE ½ measurement	D	18	7¼"	
SHOULDER	E	12	4¾"	
CUFF ½ measurement	F	11	4⅜"	
SLEEVE	G	61.5	24¼"	
FULL LENGTH AT CENTER BACK	H	54	21¼"	

FIGURE 5.52
Size chart.

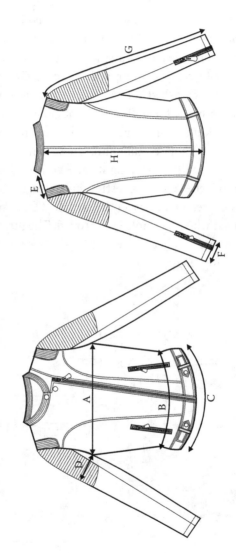

FIGURE 5.53 Measuring points.

Ladies' Tie Dress with Bodice Seams

This is another design that might have caught the eye of the buyer, which has no previous history with your company. The buyer will contact suppliers for a price and a sample. It has a simple shape, but if the bodice seam detail looks right and fits correctly and with the right choice of fabric, this could make into a very profitable commercial dress.

To keep the price down with trial runs like this, the supplier could be asked to submit suitable fabrics already available at fabric suppliers, rather than have the fabric especially made. In these circumstances, the fabric wholesaler would be asked for fabric composition, weight care instructions, and any other information that they have available. (Supplier will be asked to check if the material meets the minimum fabric performance for wovens in the supplier manual). There will probably be options of several materials in different weights and compositions. It is advisable to have one sample checked by a testing house or, if you have the facilities, check the fabric weight and wash the sample. Chart and packaging details will be finalized when a sample is approved. The buyer should tell the factory the size range that is required and if the dress could be flat packed or delivered hanging (Figure 5.54).

FIGURE 5.54
Ladies' dress with bodice seam.

- The first page of the specification, as shown in Figure 5.55, shows the overall appearance and style details of the product.
- The second page, as shown in Figure 5.56, shows the front and back details and the side zip.
- The third page, as shown in Figure 5.57, shows the size chart.
- The fourth page, as shown in Figure 5.58, shows the measuring points.

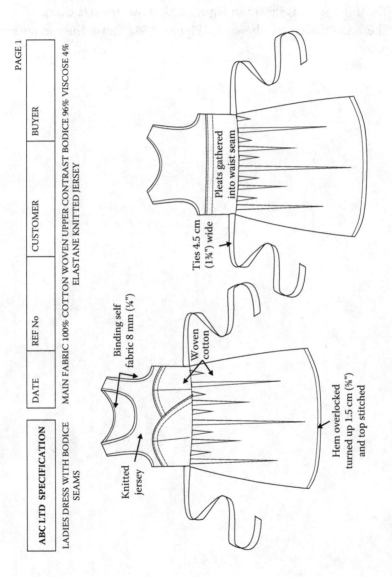

FIGURE 5.55
Overall appearance and style details of the product.

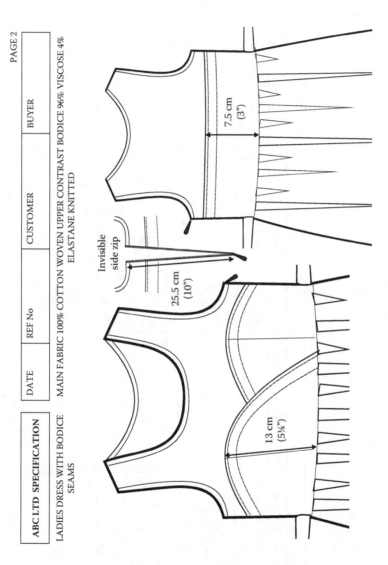

FIGURE 5.56
Front and back details and the side zip.

Size UK 12 USA 8		CM	INCHES			PAGE 3
BUST AT THE BASE OF THE ARMHOLE ½ measurement	A	42.5	16¾			
WAIST ON SEAM ½ measurement	B	42.5	16¾			
HEM ½ measurement	C	71	28			
NECK OPENING	D	21	8¼			
BACK NECK DROP	E	5	2			
FRONT NECK DROP	F	9	3½			
SHOULDER	G	7	2¾			
ARMHOLE ½ measurement	H	21.5	8½			
FULL LENGTH FROM CENTER SHOULDER	I	80	31½			
LENGTH CENTER SHOULDER TO WAIST SEAM	J	31.5	12⅜			
TIE LENGTH	K	81	31⅞			

FIGURE 5.57
Size chart.

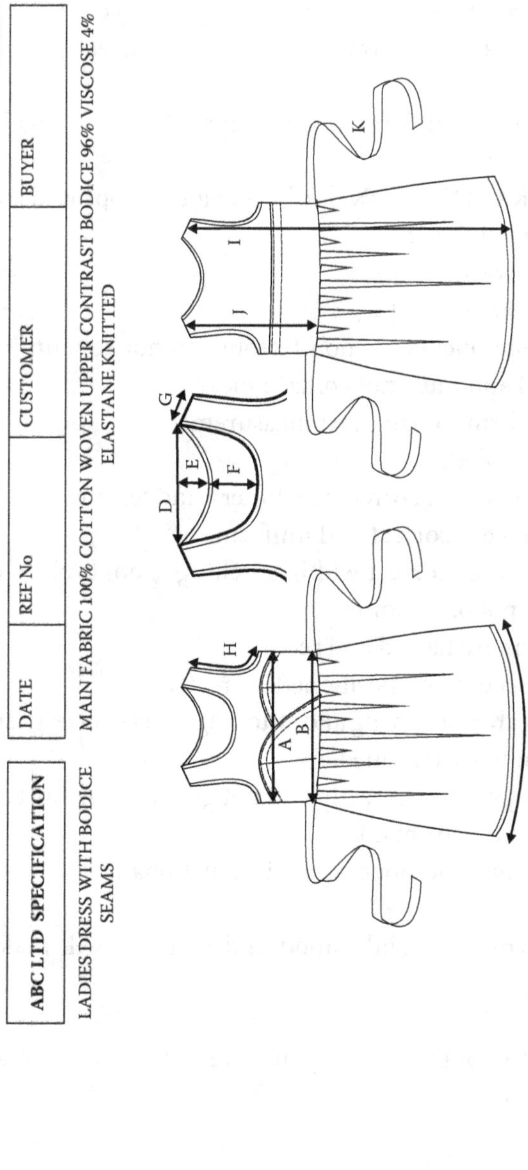
ABC LTD SPECIFICATION
DATE
REF No
CUSTOMER
BUYER
PAGE 4
LADIES DRESS WITH BODICE SEAMS
MAIN FABRIC 100% COTTON WOVEN UPPER CONTRAST BODICE 96% VISCOSE 4% ELASTANE KNITTED
A
B
C
D
E
F
G
H
I
J
K

FIGURE 5.58
Measuring points.

Sending Specifications

When sending specification to factories and requesting samples, it is a good practice to send the following reminders with the specification. It is a checklist for the factory and an opportunity to remind it about quality issues and the factory's responsibilities in the sampling procedure. The list is fairly standard but will have to be altered slightly, depending on the product.

Please ensure that all the following are checked correctly:

- Fabric weight and construction is as agreed, with no shading, flaws, or damages.
- Zips are YKK or OPTILON or of a previously approved source.
- Zips are correctly sewn.
- Colors are correct.
- Components are correct.
- All metal components are non-ferrous (do not contain iron and so will not rust) and must not contain nickel.
- Patterns conform to size chart measurements.
- Style is as specified.
- Seams are sewn with correct machinery and tension.
- Seam allowance is correct and uniform.
- Top stitching is of correct width; stitching is not broken; or rows of stitching are not badly joined.
- Ends of seams are tacked and secure.
- Buttons are secure and positioned correctly.
- Button holes are not fraying and stitching covers cut edge sufficiently.
- All cotton ends are trimmed.
- Garment are pressed correctly and not glossed or distorted.
- Garments are free from odor.
- Fabric tests must be done at an internationally accredited testing house.
- Please confirm all is understood and that you will make samples ASAP.

Do not hesitate to contact me if you have any queries or any of the information is unclear.

6
Fit and Fit Sessions

Establishing the right fit is a difficult and complex issue for companies, because no two figures are the same. Customers often wonder why they are one size in one brand but another size in a different brand; this is simply because there is no uniform sizing in the industry, and each brand will design for a particular figure's shape and height. Particularly, catalog and online companies have problems with fit because they try to develop pattern blocks that fit the majority of their customers and to minimize returns. Checks should be regularly done to look for high-returning styles, especially before repeat orders are placed. If a style has a high percentage of returns, then an investigation should follow and stock should be re-examined, because, possibly, something was missed at the fitting sessions or quality audit, and as many garments as possible should be tried on to re-assess the fit. I have known up to 50% returns on fashion items, and you cannot always trust the reasons given by the customer for returning the item. It might be fit, but the customer could just easily say that they did not like the color.

The fact that a garment measures exactly to the size chart does not mean it is acceptable for fit, because some faults do not usually show up when the garment is flat on the table, especially what we refer to as the garment balance, which indicates a problem in the fit between different pattern pieces, such as the back and the front of a garment, which can be too long or too short in comparison with the other piece. Other faults are seams not compatible, or there is not enough fullness or too much fullness, linings not fitting properly, and sleeves pitched too forward or pitched too far back; these faults show only when the garment is tried on.

Fit sessions for any new products should be an established procedure organized by the quality controller and coordinated with the buying department. As many samples as possible should be ready for fitting at the same time, as it is easier for everyone concerned to allocate time for one longer session rather than for numerous sessions spread over several days. Models used are often staff from other departments, who have to keep on top of their own work, as well as help you with the fittings, but most department managers realize that loaning their staff is benefitting the company, but a box of chocolates at Christmas will be appreciated by both the model and the manager. This reminds me the fact that you should have several models on standby, in case the one you usually employ increases in size (too many chocolates), is on sick leave, or is on holiday. Other options are to use model agencies or a size range of mannequins, which may be less hassling to organize and you won't get any comments about how bad the garment fits or how uncomfortable it is. There is really no substitute for trying garments on a body, and the ideal situation is to have in-house models, who, if necessary, can be called on at short notice. One company I worked for sold larger ladies' ranges, which offered up to a UK size 28. We relied on using previous size charts that historically had reasonable return rates, and we played safe by designing garments that were sometimes described as tents. The buyers wanted to be more adventurous and offer clothes that were more stylish but were worried about fit and high returns.

We realized that if we were to attempt this, we would have to find a larger model to try on initial samples. We did have a UK size 28 mannequin, but we were not even sure how typical this was of a size 28 customer. I did not want to wander round the head office, looking for large women and asking them what size they were! Instead, I asked one of my female colleagues to discretely look for potential models, and eventually, she found the perfect choice. The lady bought size 28, but importantly, she was not self-conscious about it and welcomed the opportunity to give feedback about not only the samples she was trying on but also what she thought in general about the clothes on offer for ladies of her size. The buyers became very involved with these sessions, as it gave them confidence to try new shapes and styles, as our model was always very vocal about what ladies of her size were comfortable wearing.

With fit being so important, many companies employ pattern cutters, who create pattern blocks to send to their factories. Other companies think that the best solution for the factories is to develop their own pattern blocks, or often, it could be a combination of the two methods. As a final note, where possible suppliers or their agents should also attend the fit sessions, it's much better if there is a problem and they can see it first hand for themselves. Here, skills in time management come in to play, as there are so many people involved. Whichever route you take, there is no guarantee as to which garments your customers will like and which will sell with low returns, but fitting sessions are a very important tool to help keep your company's share of the market.

7
Fabric Specification and Performance

Technologists and technical designers from a pattern and sewing background will probably have a basic knowledge of fabric construction and properties. If the fabric quality and testing come under your responsibility, you need to know about setting standards and the testing that should be done.

Setting the right quality for fabrics is a very important part of range building. Setting the quality too high and using fabrics that can account for at least 50% of the overall garment cost will be too expensive. Set the standard too low and you will risk a high percentage of complaints and lose customer loyalty.

Our first test of a fabric is often to check if we like the feel, the handle, and the appearance and to see how it will look and drape in the finished garment. A good-quality fabric from natural fibers such as wool, silk, linen, and cotton will usually always feel and drape well, but price might be prohibitive for your target market, and they might only be dry cleaned, which then limits the customers you can sell to. There are many options of synthetic or blends of synthetic and natural fibers, which are suitable cheaper alternatives. Once we have found a suitable base fabric, we need to establish that the garment will not fall into pieces when worn or when put through the laundering process. This may sound extreme, but I have seen seams coming apart with a simple tug or fabric so badly constructed that it tears under slight pressure. Minimum fabric performances have now been adopted by most companies and are an established procedure for selecting fabrics.

The fabric performance specification refers to the durability of the fabric when made into a garment; it defines any changes to the fabric during washing or dry cleaning and how robust the fabric and seams will be during wear. The type of fabric we choose is also determined by the end use.

For example, school clothing and work wear need fabric and seams that are strong and stand up to continuous hard wear and therefore the yarn needs to be strong and the warp and weft weaved tightly together. A fashion wool jacket is likely to be the opposite in structure, designed to have an open weave and raised surface detail and to be classified more as a decorative fabric and would not be a hard wear.

There are no specific international standards for any fabric quality or performance, as this is for you to decide what is suitable for your products, but there are recognized adopted standards for general apparel. Internationally recognized tests refer to the method of checking that the fabrics meet your requirements.

Factories specialize in particular products. For example, a trouser manufacturer will have a range of different fabrics suitable for trousers to show a buyer from mills that they regularly source from. A good supplier will know the quality of the fabrics and help the buyer choose what is suitable for their business, and many buyers gain extensive knowledge on fabrics for ranges that they specialize in buying. That said, it is still necessary to record the fabric details and have it tested to see that it meets your required standard and the fiber composition and construction are as declared.

When a new fabric has been selected, the mill or factory should supply you with as much technical information as possible about the fabric construction, composition, stability to washing, and dye fastness, as these are formulated during the making and finishing of the fabric.

It is important that the suppliers commit themselves to the quality of the fabric that they intend to supply. The supplier has to be specific about the technical details, which should always be readily available. If we don't insist on specifics, we could be supplied with a poorer variation of what we expected. For instance, if you are buying a poly/cotton blend, there are many different options, and you may start with a 75% cotton and 25% polyester blend and eventually finish with a 50/50 blend. Suppliers, if possible, try to keep their options open, and if, for example, the price of cotton goes up, they could take the easy option and buy an alternative blend with more polyester. In some instances, you may have no choice, but the supplier has to be upfront with you and explain why he wants to substitute a different fabric and send to you the new fabric details. The new details from the supplier will then be declared to the laboratory when the fabric is sent for testing. The supplier needs to be aware that you are constantly monitoring all fabric and trims and that nothing can be substituted without your approval.

Start as you mean to go on, you must be in control of your production, and letting the supplier make substitutes without your knowledge is a slippery slope to poor quality.

Suppliers should be encouraged to compile and send you a sample book with all examples of locally obtainable fabrics that they use or are likely to use, giving as much technical detail as possible, including the following:

- Fiber composition
- Weight
- Construction
- Yarn count

The buyer, where possible, can then quote a fabric quality when selecting a range or a garment, enabling the supplier to give a more accurate costing when preparing to make samples.

For examples of fabric construction, see fabric specification for the men's short-sleeved shirt and ladies' fashion trouser in Chapter 5. For fabric's minimum performance specifications, see Chapter 9. Although one fabric is woven and one knitted, the tests are similar.

First, the composition and construction for the base fabric are specified and laboratories will analyze the fiber composition and check the yarn count and fabric weight. If these are not correct, then there is little point of continuing testing.

The second part of the specification concerns the fabric performance. It is usual to state what test you want and then the method and the result you require.

Finally, we state the suggested laundering instructions, and the lab will use those for testing stability and color fastness.

You may have a number of garments styles with the same base fabric and performance standards or with different base fabrics, where the same performance standards will be applicable. Over a period of time, you will create a library of core proven fabric specifications that buyers will take with them when visiting suppliers sourcing new styles, and they will be able to give the factory either the exact fabric specification or one very similar, which will save much time further down the supply chain.

8
Fabric Testing

It is necessary to validate the information that the mill or manufacturer gives you. This is normally done by having fabric tested by an accredited testing house (internationally recognized) either in the country of manufacturer or locally in your head office. Some companies have their own equipment and employ qualified fabric technicians to carry out the testing at their headquarters. When using an accredited testing house, it will not matter in which country the test is done, because the test has to be carried out to the agreed method, and no matter which country does the test, it will be done in exactly the same way. As the fabric construction is such an important part of the garment, when finalized, the fiber composition, weight, and yarn count should be added to your contract with the supplier.

The testing house will give you guidance on the result you should be achieving and the method that you should be using.

The following are brief descriptions of the types of tests done to establish if the fabric is fit for purpose.

- *Fabric construction*: You have a legal responsibility to ensure that the fiber composition label on the garment is correct. You also need to know that the supplier has not substituted an alternative fabric or blend without informing you. Fiber construction (number of warp and weft threads per inch or the machine gauge and number of threads for knitted garments) and yarn count are the components that go into making of the base fabric.
- *Fabric weight*: Fabric weight is a critical factor; we may choose a heavier quality to keep us warm or to make the garment more hard wearing. The designer may simply select a heavyweight fabric, because it suits the style better. Often, the same style of garment is offered in summer and winter weights. Once a suitable-weight

fabric is selected, we expect no more variation than ±5% for the production fabric.
- *Seam strength*: This test method is used to determine the maximum force for sewn seams, when the force is applied perpendicularly to the seam. The test can be applied to fabric samples sewn to form a seam or the production seam, as received in finished garments. The major contributors to seam strength are fabric type and weight, thread type and size, stitch and seam construction, stitches per inch, and stitch balance.
- *Seam slippage*: This test method is used to determine the resistance to slippage of the warp and weft yarns, using a standard seam. It is used as an indication of the tendency of yarns to slip at a seam when stress is applied. The result is that the yarns pull out, but the stitch doesn't rupture.
- *Tear strength*: The test reflects the strength of the yarns, fiber bonds, and fiber interlocks, as well as their resistance to tearing.
- *Abrasion and pilling*: These two tests measure the amount of surface wear of a fabric. Pilling, fuzzing, color change, and the breaking of fibers are unsightly on a garment, and the tests set a minimum standard of rub before there is a noticeable change in the appearance of the fabric surface.
- *Stability to washing and dry cleaning*: These tests check the relaxation of the fabric after laundering by following the garment care label. Fabric can spread out as well as shrink, so normally, a tolerance of ±3% in the width and length for woven fabrics and 5% for knitted fabric is acceptable.
- *Color fastness*: Color fastness can be a serious problem, especially with dark or deep colors. It is the property of a dye or print that enables a fabric to retain its depth and shade throughout the wear life of a product. Dyes are considered fast when they resist deteriorating influences such as laundering or dry cleaning. Fabric can lose color with the use of detergent solution and abrasive action during hand or machine washing and stain other garments with which it is being washed at the same time. Dye fastness to washing is tested by attaching the fabric to a piece of multi-fiber strip containing bands of acetate, cotton, nylon, polyester, acrylic, and wool during a washing process. Other tests called wet and dry rub check for loss of dye and staining when the garment is worn; if the amount of dye coming off the fabric is excessive, then potentially,

it can rub off onto other garments that the customer is wearing or even onto light-colored furniture. Further tests such as dye fastness to perspiration and to chlorine for swimwear can be done.

I have stressed the importance of fabric testing as being an essential procedure in quality assurance, and using accredited testing houses is an expensive but absolute necessity, unless you have your own qualified textile technicians and equipment. However, it is possible with a modest outlay and basic training to do some testing yourself. This is no substitute for using professional testing houses, but it can give you the opportunity, if necessary, to check quickly and inexpensively the basic properties of new fabric submitted by suppliers.

1. *Fabric weight, requirement circular fabric cutter, and digital scales*: This enables you to take a sample of fabric, and when weighed, it gives you the weight in grams per square meter.
2. *Stability to washing (requirement—domestic washing machine)*: This enables you to wash fabric or garment as per care label and check the shrinkage.
3. *Dye fastness, requirement crock meter, white cotton squares, multi-fiber strips, and a gray scale*: This enables you to check the dye fastness to washing and the wet and dry rub, measuring the results with a gray scale. The scale consists of nine pairs of gray-colored chips, each representing a visual difference and contrast.

A reading of 5 means no change and 1 means the worst transference of color.

Doing these tests yourself is not an alternative to having them done at a test house, but with basic training, you will become experienced at achieving a reasonably accurate result.

9

Supplier's Manual

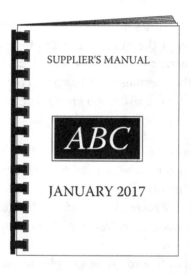

Sometimes called the company's bible, the manual will contain terms and conditions for trading with their suppliers. It is essential that before doing business with your company, the supplier is fully aware of your standards and method of working, and often the manual will include an acceptance form to be signed by a director of the manufacturing company. Manufacturers often work with many customers in different countries, and it would be ideal for them if they all accepted the same quality standards and procedures. However, all companies are different, with diverse customer requirements and warehouses that operate differently. When talking to suppliers, you should always reinforce what you have requested in the manual and explain why the standards and procedures are so important for your company to be successful, which could result in more business being placed with their factory. These are usually the areas covered in the manual, and you can appreciate that it is an important tool for quality assurance and a source of reference for everyone in your company as well as for suppliers. Usually, the quality controller will have the responsibility for compiling the content, but all departments should have the opportunity to contribute.

SAMPLING PROCEDURES

This explains to your factories the stages required by your company for the successful completion of the manufacturing process, from first samples to final delivery. It clearly shows that the manufacturer cannot proceed to the next stage until the previous stage has been completed and signed off. Preach the message that the risk is too great if corners are cut or shortcuts taken by not completing each stage correctly.

Stage 1: First approval samples
   a. Minimum two pieces (sizes as requested by the quality control [QC] department).
   b. Submitted in the correct base fabric, with lab dips for color approval.
   c. Trims and accessories representative of production, where possible.
   d. *When approved?*

Stage 2: Approval of fit and grade samples in correct base fabric
   a. Submit samples in sizes as agreed with QC, with comments.
   b. *When approved, final chart, label, and packaging instructions will be issued.*

Stage 3: Preproduction
   a. Supplier to send laboratory fabric-approval certificate and laundering-test-approval certificate.
   b. Four preproduction samples across the size range with a sample report and complete with packaging and labeling.
   c. *When approved?*

Stage 4: Production
   a. Four production samples across the size range with the sample report.

b. Required 10 working days before shipment day. (see Chapter 11, which shows the timeline for sampling, production, and delivery.)

FABRIC MINIMUM PERFORMANCE STANDARDS

Suppliers should always be reminded of the fabric standards that you and your customers expect and the quality levels that have made your business successful. When buyers source from existing and new suppliers, they should always discuss fabric standards at the start of the selection, so as to avoid any misunderstandings at a later stage. If a fabric is not acceptable because the supplier was not aware of your performance standards, a higher price might have to be negotiated.

Minimum fabric performance details are ideally suited to go into the supplier's manual as well as being sent to manufacturers with each specification, as you should always take the opportunity to remind your supplier of your standard requirements.

Tables 9.1, 9.2, and 9.3 are examples of typical retail standards under the headings of jersey wear, knitwear, and wovens and refer to any type of garment that comes under each of these generic headings.

The first column, heading test, will tell you all the important things you want to know about the fabric—how it will stand up to wear and laundering.

The second column is the method used to measure the test results, and these are especially *set in stone*. In a world of commercial uncertainty, these methods are internationally recognized, and if you use testing houses that are accredited and approved to test these methods, you will know in whichever country your fabric is tested, the results will be accurate. These tests should never be confused with wear testing, which is a totally separate procedure. Sometimes, it might be considered useful to distribute garments out to various people, usually in head office, which they will wear for a period of time and launder them as part of their normal wash load, and the garments would then be assessed. This does have a certain value, and sometimes, you can get useful feedback, but it is not methodical and controlled as a lab test and can throw up anomalies.

The third column declares your minimum requirements after completion of the test. These do not have to be set in stone; they might be better than your minimum requirements, marvellous!! Or the tests might show that they do not meet your standard. This does not mean automatic rejection; it means that you have to assess the results and make a decision, which first depends on whether the fabric is a first sample. You can then afford to reject it, sending the result to the supplier and telling them to resubmit correct fabric. If the fabric is from bulk and in the factory, waiting to be cut, you have to make a "commercial decision," depending on the price and end use of the garment. This is always difficult, as pressure will always be on you to accept to avoid delay in manufacturing and eventually loss of sales. Let's take an example of stability to washing. We require ±3% and the fabric shrinks 10%, no problem! We can confidently reject it with no come back on your decision. But if the shrinkage is, for example, 4%–6%, can we accept this result? It might have occurred to the reader that we have set the minimum standards slightly higher and that 5% would be acceptable, but to make sure we achieve that, we set the standard at 3%, and if we had set the standard at 5%, we are more likely to get 6% or 7%.

Although much thought is needed when setting minimum standards, you will also realize that to achieve consistent quality, your choice of factories and choice of fabric suppliers will play a big role. The more direct involvement you can have in controlling your fabric source, the fewer problems you will have in the long term. Making decisions on border line results is never straightforward, and as with all aspects of quality assurance, you will gain knowledge and experience as to when and where to draw the line, and testing houses you work with will give advice on test results that don't meet your requirements.

The fourth column, comments, is an add-on and is used where clarification might be needed on the testing (Table 9.1).

- The second page, as shown in Table 9.2, shows Intertek knitwear standards.
- The third page, as shown in Table 9.3, shows Intertek woven standards.

TABLE 9.1
Intertek's Jersey Standards

	Performance Standard Jersey Wear		
Test	Method	Requirements	Comments
Mandatory Tests			
Fiber composition	EU Regulation 1007/2011/EU	No tolerance for 100% composition +/− 3% on blends	
Chemical hazards	R.E.A.C.H	Compliance	
Base Test Requirements: All Physical Tests to Be Completed on One Color Way			
Stability to washing	BS EN ISO 6330: 2012	Knitted +/− 5%	Test at care label temperature
Stability to dry clean	Commercial	Knitted +/− 5%	Dry clean items only
Spirality	EN 16322-2 Procedure C	5% Max	Single jersey weft knits only
Bursting strength	BS EN ISO 13938-1 or 2: 1999	180 kPa Lace 250 kPa	
Martindale pilling	BSEN ISO 12945-2: 2000	Knitted 500 Revs Grade 4	Spun synthetics, wool, acrylic and their blends
Fabric weight	BS EN 12127: 1998	+/− 5% of claimed weight	As buyers agreed/approved weight
Stretch and recovery (After 1 and 30 mins)	BS EN 14704-1: 2005 Method A	10% Max	
Snagging resistance— pod Method	BS 8479: 2008	Grade 4	Fabrics containing filament yarns only
Bulk Test Requirements: Color Fastness to Be Completed on All Color Ways. Dimensional Stability to Be Completed on All Color Ways			
Stability to washing	BS EN ISO 6330: 2012	Knitted +/− 5%	Test as care label temperature
Stability to dry clean	Commercial	Knitted +/− 5%	Dry clean items only
Spirality	ISO 16322-2 Procedure C	5% Max	Single jersey weft knits only
Color fastness to washing	BS EN ISO 105-C06: 2010	Change: 4 Stain: 4 Cross Stain: 4/5	Test as care label temperature
Color fastness to dry cleaning	BS EN ISO 105-D01: 2010	Change: 4 Stain: 4 Cross Stain: 4/5	Dry clean items only
Color fastness to water	BS EN ISO 105-E01: 2013	Change: 4 Stain: 4 Cross Stain: 4/5	N/A on white and cream fabrics
Color fastness to perspiration	BS EN ISO 105-E04: 2013	Change: 4 Stain: 4 Cross Stain: 4/5	Must be tested on all color ways
Color fastness to rubbing	BS EN ISO 105-X12: 2002	Dry: 4 Wet: 3–4	N/A on white and cream fabrics
Color fastness to light	BS EN ISO 105-B02: 2013	Std: 4	

TABLE 9.2

Intertek Knitwear Standards

	Performance Standard Knitwear		
Test	Method	Requirements	Comments
Mandatory Tests			
Fiber composition	EU regulation 1007/2011/EU	No tolerance for 100% composition +/- 3% on blends	
Chemical hazards	R.E.A.C.H	Compliance	
Base Test Requirements: All Physical Tests to Be Completed on One Color Way			
Stability to washing	BS EN ISO 6330: 2012	+/- 5%	Test at care label temperature
Stability to dry clean	Commercial	+/- 3%	Dry clean items only
Martindale pilling	BS EN ISO 12945-2: 2000	Knitted 500 Revs Grade 4 After Cleanse	Spun synthetics, wool, acrylic and their blends
Fabric weight	BS EN 12127: 1998	+/- 5% of claimed weight	As buyers agreed/ approved weight
Bulk Test Requirements: Color Fastness to Be Completed on All Color Ways. Dimensional Stability to Be Completed on All Color Ways			
Stability to washing	BS EN ISO 6330: 2012	+/- 5%	Test as care label temperature
Stability to dry clean	Commercial	+/- 3%	Dry clean only items
Color fastness to washing	BS EN ISO 105-C06: 2010	Change: 4 Stain: 4 Cross Stain: 4/5	Test as care label temperature
Color fastness to dry clean	BS EN ISO 105-D01: 2010	Change: 4 Stain: 4 Cross Stain: 4/5	Dry clean items only
Color fastness to water	BS EN ISO 105-E01: 2013	Change: 4 Stain: 4 Cross Stain: 4/5	N/A on white and cream fabrics
Color fastness to perspiration	BS EN ISO 105-E04: 2013	Change: 4 Stain: 4 Cross Stain: 4/5	To be carried out on all colorways
Color fastness to rubbing	BS EN ISO 105-X12: 2002	Dry: 4 Wet: 3–4	N/A on white and cream fabrics
Color fastness to light	BS EN ISO 105-B02: 2013	Std: 4	

TABLE 9.3

Intertek Woven Standards

	Performance Standard Woven		
Test	Method	Requirements	Comments
Mandatory Tests			
Fiber composition	Eu regulation 1007/2011/EU	No tolerance for 100% composition +/– 3% on blends	To be completed on all base material qualities once per season. 100% may have letter of certification from supplier
Physical Test Requirements			
Stability to washing	BS EN ISO 6330: 2012	+/– 3%	Test at care label temperature
Stability to dry clean	Commercial	+/– 3%	Dry clean items only
Tensile strength	BS EN ISO 13934-2: 2014 – Grab Method	150N	
Seam slippage	BS EN ISO 13936-1: 2004	6 mm SO 80N	Seam slippage to be tested at 3 mm, on fabrics with contrast warp and weft threads
Seam strength	BS EN ISO 13935-2: 1999	SS 120N	
Tear strength	BS EN 13937-1: 2000	10N	
Martindale abrasion	BS EN ISO 12947-2: 1999	S/C at 5000 Revs Grade 3/4	10,000 Revs delicates 20,000 Revs all others
Martindale pilling	BS EN ISO 12945-2: 2000	2,000 Revs Grade 3/4	Only test on spun synthetics, wool, acrylic and their blends
Stretch and recovery (After 1 min)	BS EN 17404-1: 2005 Method A (30N Load)	5% Max	Fabrics containing less than 6% elastane
Color Fastness and Appearance Test Requirements			
Color fastness to washing	BS EN ISO 105-C06: 2010	Change: 4 Stain: 4 Cross Stain: 4/5	Test as care label temperature
Color fastness to dry clean	BS EN ISO 105-D01: 2010	Change: 4 Stain: 4 Cross Stain: 4/5	Dry clean items only
Color fastness to water	BS EN ISO 105-E01: 2013	Change: 4 Stain: 4 Cross Stain: 4/5	N/A on white and cream fabrics
Color fastness to rubbing	BS EN ISO 105-X12: 2002	Dry: 4 Wet: 3–4	N/A on white and cream fabrics
Color fastness to light	BS EN ISO 105-B02: 2013	Std: 4 Req: 4	
Print durability	IHTM-01/01B	Change: 4 – N.S.C. Cross Stain: 4/5	Test at care label temperature Refer to method for other requirements
Appearance after wash	IHTM-07	Refer to method	Test at care label temperature

BASIC SIZE CHARTS AND SPECIFICATIONS FOR CORE LINES

Some styles your company offer will change very little over the seasons and are referred to as core basic lines, selling in high volume, where you have achieved a good fit with low returns. The first example is a men's shirt. The style of the collars, cuffs, and pockets may alter, but these are all superficial changes. With this type of product, it is a good idea to create a template to use as a basis for all shirts. I have created a file that shows a range of shirts, formal and casual, that a company might want to offer, giving each a style reference. Separately, I have shown a selection of collars, pockets, and cuffs that are interchangeable with any of the body shapes. As these are established products, the size chart is complete, and in this format, it can be circulated to new and existing factories. I have also included the basic outline measurements for a ladies' A-line skirt and shirt blouse, which can be adapted for many other styles.

These and similar products are ideal to be included in the manual, as they are proven charts and a good starting point for many other styles. Manufacturers should always be encouraged to sample new styles and fabrics that the buyers might be interested in but should always refer to the manual for guidance on your standards first (Figure 9.1).

- Second page, as shown in Figure 9.2, shows style F2 tapered-fit shirt, long sleeve.
- Third page, as shown in Figure 9.3, shows style F3 standard-fit shirt, short sleeve.
- Fourth page, as shown in Figure 9.4, shows style C1 standard-fit casual shirt, long sleeve.
- Fifth page, as shown in Figure 9.5, shows style C2 tapered-fit casual shirt, long sleeve.
- Sixth page, as shown in Figure 9.6, shows style C3 standard-fit casual shirt, short sleeve.
- Seventh page, as shown in Figure 9.7, shows pocket styles.
- Eighth page, as shown in Figure 9.8, shows cuff styles.
- Ninth page, as shown in Figure 9.9, shows collar styles.
- Tenth page, as shown in Figure 9.10, shows position of labels.

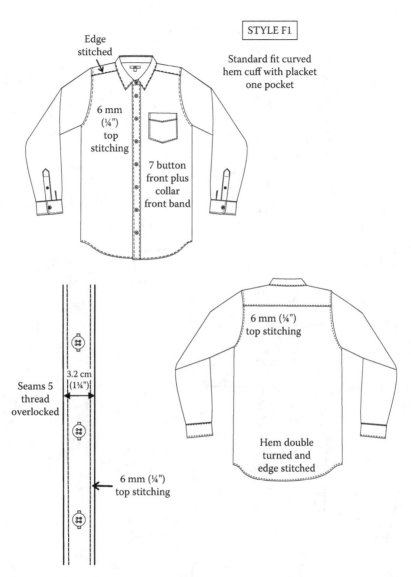

FIGURE 9.1
Style F1 standard-fit shirt, long sleeve.

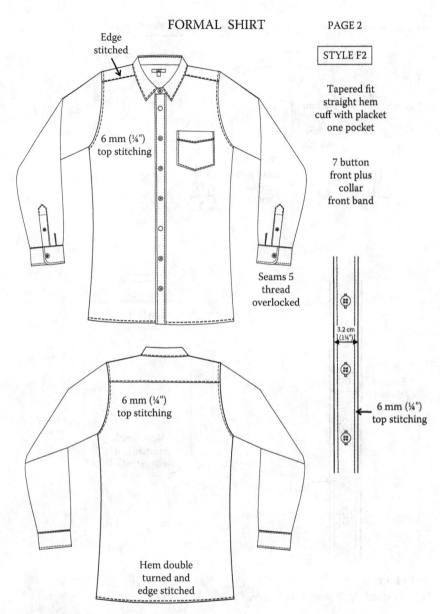

FIGURE 9.2
Style F2 tapered fit shirt long sleeve.

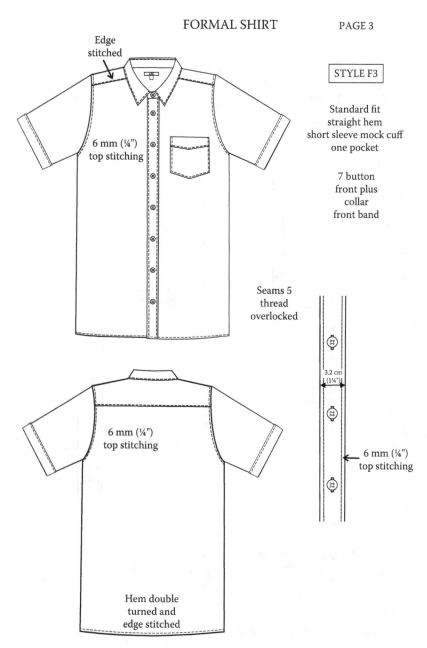

FIGURE 9.3
Style F3 standard fit shirt short sleeve.

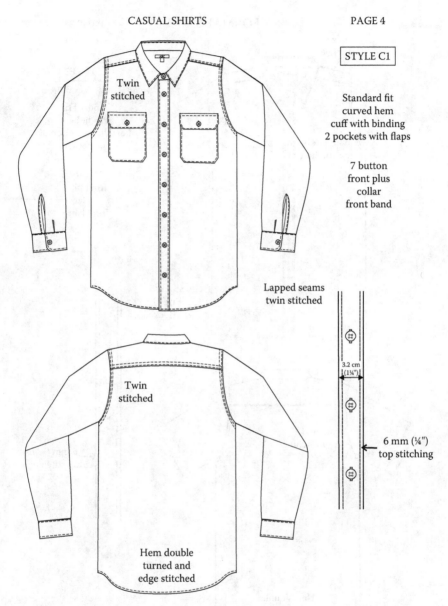

FIGURE 9.4
Style C1 standard fit casual shirt long sleeve.

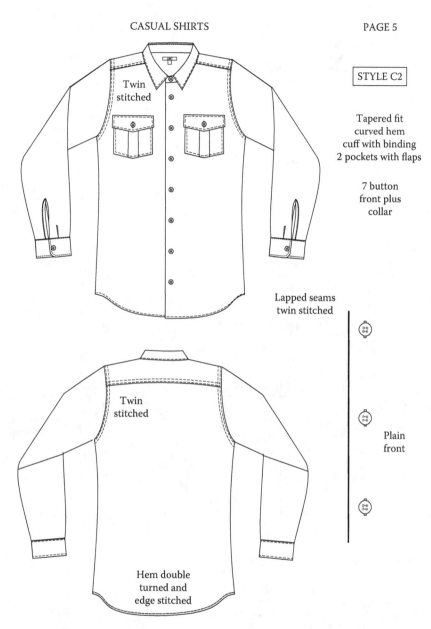

FIGURE 9.5
Style C2 tapered casual shirt long sleeve.

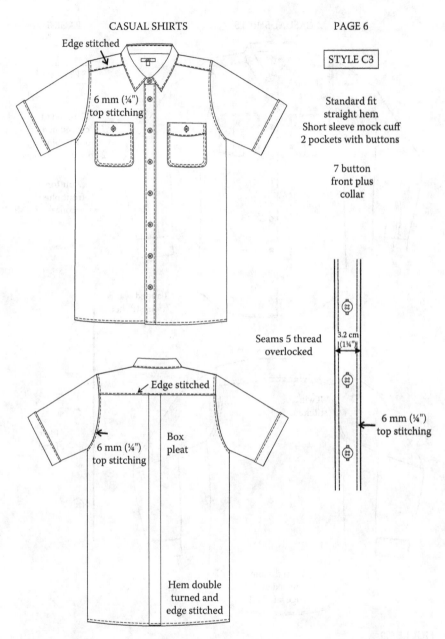

FIGURE 9.6
Style C3 standard fit casual shirt short sleeve.

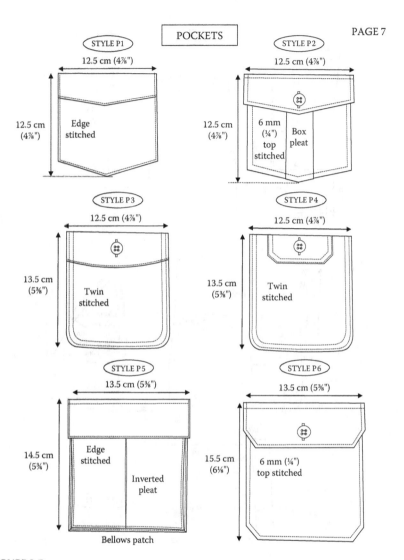

FIGURE 9.7
Pocket styles.

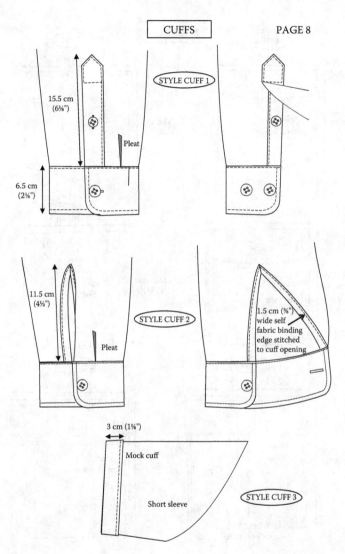

FIGURE 9.8
Cuff styles.

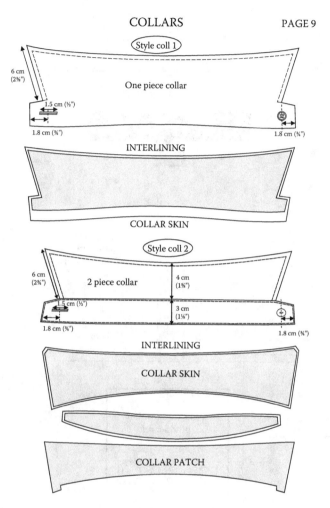

FIGURE 9.9
Collar styles.

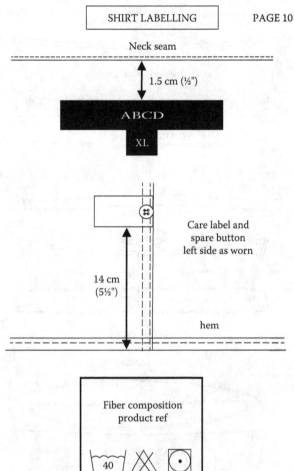

FIGURE 9.10
Position of labels.

At a first glance, the following size chart may look complicated, but view it as a master template covering a range of men's shirts in all sizes and style options. Sometimes, it is helpful to limit what you offer if for no other reason but to reduce your stock holding. Often, styles are added to a range when there is very little difference in the measurements. This chart has rationalized the body shapes to two, with the option of long and short sleeves. The shaded columns indicates the different sizes for the casual shirt, S to XXXL, and the collar size on which each is based on.

The chart based on the best-selling lines with lowest return rates standardizes measurements for future production (Figure 9.11).

- Second page, as shown in Figure 9.12, shows the measuring points.

Size	Chest	Waist easy fit	Waist fitted	Hem	Back yoke	Bicep	Cuff opening long sleeve fastened ½ measure	Cuff opening short sleeve ½ measure	Center back length	Long sleeve length	Short sleeve length
14½" collar inches	42⅞"	42⅞"	39¾"	42⅞"	18⅜"	16⅛"	4⅜"	6¼"	29⅞"	24¾"	10¼"
14½" collar cms	109	109	100	109	46.6	41	11	16	76	63	26
SMALL inches	42⅞"	42⅞"	39¾"	42⅞"	18⅜"	16⅛"	4⅜"	6¼"	29⅞"	24¾"	10¼"
SMALL cms	109	109	100	109	46.6	41	11	16	76	63	26
15" collar inches	44⅞"	44⅞"	41½"	44⅞"	18⅞"	16¾"	4⅜"	6¼"	29⅞"	24⅞"	10¼"
15" collar cms	114	114	105	114	47.8	42.5	11	16	76	64	26
15½" collar inches	46⅞"	46⅞"	43½"	46⅞"	19⅜"	17¼"	4¾"	6⅝"	31"	25⅛"	10½"
15½" collar cms	119	119	110	119	49	44	12	17	79	64	27
MEDIUM inches	46⅞"	46⅞"	43½"	46⅞"	19⅜"	17¼"	4¾"	6⅝"	31"	25⅛"	10½"
MEDIUM cms	119	119	110	119	49	44	12	17	79	64	27
16" collar inches	48⅞"	48⅞"	45½"	48⅞"	19¾"	17⅞"	4¾"	6⅝"	31"	25½"	10½"
16" collar cms	124	124	115	124	50.2	45.5	12	17	79	65	27
16½" collar inches	50⅞"	50⅞"	47½"	50⅞"	20¼"	18½"	4¾"	7⅞"	31⅞"	25½"	11"
16½" collar cms	129	129	120	129	51.4	47	12	18	81	65	28
LARGE inches	50⅞"	50⅞"	47½"	50⅞"	20¼"	18½"	4¾"	7⅞"	31⅞"	25½"	11"
LARGE cms	129	129	120	129	51.4	47	12	18	81	65	28
17" collar inches	52¾"	52¾"	49½"	52¾"	20¾"	19"	5"	7⅞"	33½"	26"	11"
17" collar cms	134	134	125	134	52.6	48.5	13	18	85	66	28
17½" collar inches	54¾"	54¾"	51½"	54¾"	21⅛"	19½"	5"	7½"	33½"	26"	11½"
17½" collar cms	139	139	132	139	53.8	50	13	19	85	66	29
XL inches	54¾"	54¾"	51½"	54¾"	21⅛"	19½"	5"	7½"	33½"	26"	11½"
XL cms	139	139	132	139	53.8	50	13	19	85	66	29
18" collar inches	56¾"	56¾"	53½"	56¾"	21½"	20¼"	5"	7½"	34¼"	26"	11½"
18" collar cms	144	144	137	144	55	51.5	13	19	87	66	29
18½" collar inches	58¾"	58¾"	56¾"	58¾"	22⅛"	20⅞"	5½"	7⅞"	35"	26⅜"	11⅞"
18½" collar cms	149	149	144	149	56.2	53	14	20	89	67	30
XXL inches	58¾"	58¾"	56¾"	58¾"	22⅛"	20⅞"	5½"	7⅞"	35"	26⅜"	11⅞"
XXL cms	149	149	144	149	56.2	53	14	20	89	67	30
19" collar inches	60¾"	60¾"	58¾"	60¾"	22½"	21½"	5½"	7⅞"	35½"	26⅜"	11⅞"
19" collar cms	154	154	149	154	57.4	54.5	14	20	90	67	30
19½" collar inches	62¾"	62¾"	60¾"	62¾"	23¾"	22"	5⅞"	8¼"	35½"	26¾"	12¼"
19½" collar cms	159	159	154	159	58.6	56	15	21	90	68	31
XXXL inches	62¾"	62¾"	60¾"	62¾"	23¾"	22"	5⅞"	8¼"	35½"	26¾"	12¼"
XXXL cms	159	159	154	159	58.6	56	15	21	90	68	31

FIGURE 9.11
Excel chart.

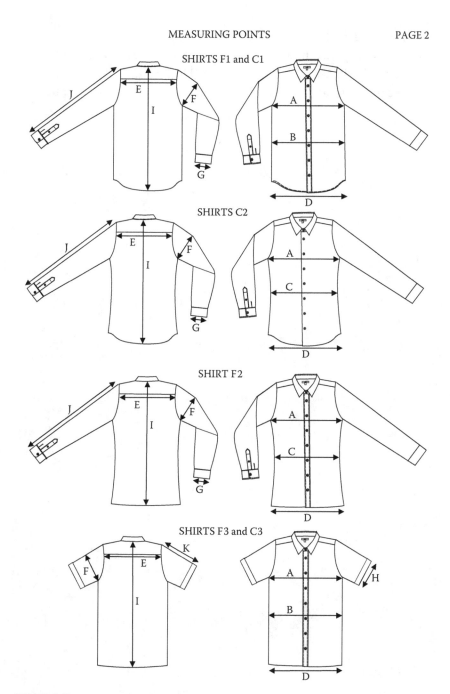

FIGURE 9.12
Measuring points.

- Ladies basic shirt's size chart (Figure 9.13).
- Second page, as shown in Figure 9.14, shows measuring points.
- Ladies basic A-line skirt's size chart (Figure 9.15).
- Second page, as shown in Figure 9.16, shows measuring points.

LADIES BASIC SHIRT

		8		10		12		14		16		18	
FULL MEASUREMENTS		cms	inches	cms	inches	cms	inches	cms	inches	cms	inches	cms	inches
BUST 2. 5 CMS BELOW ARMHOLE	A	87	38¼"	92	40¼"	97	42¼"	102	44¼"	107	46¼"	113	48½"
WAIST	B	82	32¼"	87	34¼"	92	36¼"	97	38¼	102	40¼"	108	42½"
HEM	C	94	37"	99	39"	104	41"	109	43"	114	45"	120	47¼"
SHOULDER	D	11.9	4⅝"	12.2	4¾"	12.5	4⅞"	12.8	5"	13.1	5⅛"	13.4	5⅜"
BICEP 2.5CMS BELOW ARMHOLE	E	31.5	12⅜"	32.75	12¾"	34	13⅛"	35.25	13½"	36.5	13⅞"	37.75	14½"
BACK WIDTH 12CMS FROM NECK SEAM	F	35.5	14"	36.75	14⅜"	38	14¾"	39.25	15⅛"	40.5	15½"	41.75	16⅛"
CUFF	G	21	8¼"	21.5	8⅜"	22	8½"	22.5	8⅝"	23	8¾"	23.5	9⅛"
OVERARM	H	59	23¼"	59.5	23⅜"	60	23⅝"	60.5	23⅞"	61	24	61.5	24¼"
CENTER BACK LENGTH	I	64	25¼"	64	25¼"	64	25¼"	64	25¼"	65	25⅝"	65	25⅝"

FIGURE 9.13
Size chart.

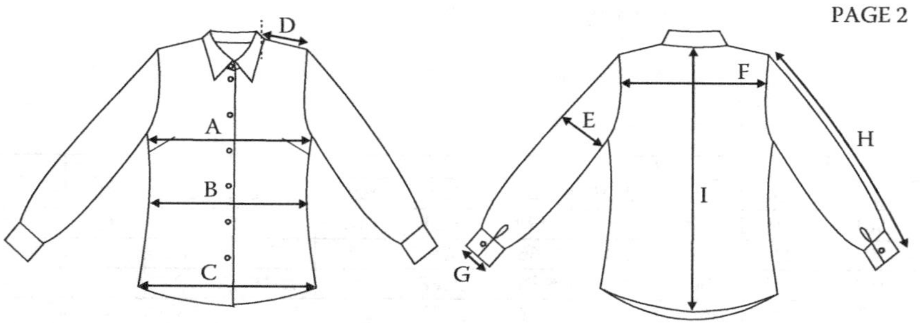

FIGURE 9.14
Measuring points.

LADIES BASIC A LINE SKIRT

		8		10		12		14		16		18	
FULL MEASUREMENTS		cms	inches	cms	inches	cms	inches	cms	inches	cms	inches	cms	inches
WAIST	A	64	25¼"	69	27¼"	74	29¼"	79	31¼"	84	33¼"	91.5	36"
MID HIP 10 CM FROM TOP EDGE	B	81	31⅞"	86	33⅞"	91	35⅞"	96	37⅞"	101	39⅞	108.5	42⅞"
FULL HIP 20 CM FROM THE TOP EDGE	C	89	35"	94	37"	99	39"	104	41"	109	43"	118.5	46⅝"
HEM	D	115	45⅜"	120	47⅜"	125	49⅜"	130	51⅜"	135	53⅜"	140	55⅛"
LENGTH FROM TOP EDGE	E	67	26⅜"	67	26⅜"	67	26⅜"	67	26⅜"	67	26⅜"	67	26⅜"

FIGURE 9.15
Size chart.

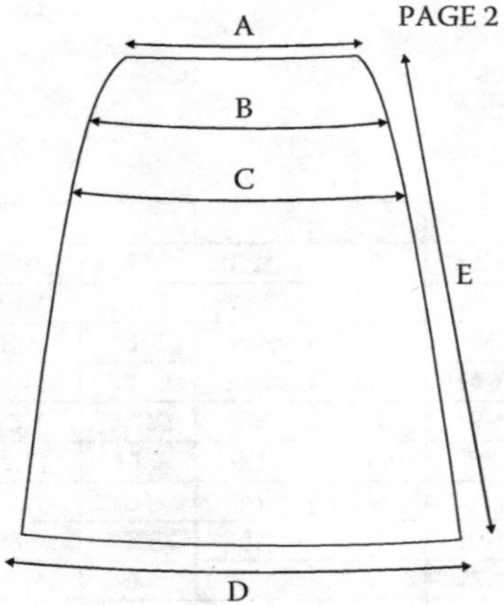

FIGURE 9.16
Measuring points.

The mantra of quality should be repeated at every opportunity, so the following should be included.

MANUFACTURING GUIDELINES

Please ensure the following:

- Fabric is of the correct quality and shade.
- All style details are correct.
- Checks and stripes match as agreed.
- Interlinings and/or sewn in correctly.
- Seams are sewn with the correct tension and there is no puckering.
- Top stitching is uniform and even.
- All seams are secured; zips and pockets are tacked where necessary.
- When zips are inserted, the seam is not stretched.
- Button holes are cut, trimmed, and sewn properly.
- Buttons are securely sewn.
- Press studs are securely attached.
- Linings fit correctly and are not distorted, twisted, or tight.
- Pockets are level.
- Measurements are to the size chart.
- Garments are free of dirty marks and stains.
- Labeling is as specified.
- Care must be taken during pressing to avoid press marks and distortion.
- Knitted garments *must not* be stretched to achieve the correct dimensions.
- Packaging and presentation are correct.
- Garments have a final inspection before shipping.

PACKAGING AND PRESENTATION

The importance of presentation should never be underestimated, as retailers put a great deal of importance on how the products look at the point of sale; in fact, it could be considered equally as important as the quality of the product, and a lot of time and money are spent on designing packaging so that it is attractive and appealing to the customer.

Every item on display should be identical, like soldiers lined up on parade. It looks unprofessional and messy if each item has tickets or labels positioned differently or the packaging is too big or small for the product. An untidy display will put the customers off from buying, as they will judge the product by its cover, and packaging samples should be processed in the same way as the product itself.

Factories do source packaging, but most retailers source their own labels and packaging through print companies that have offices in Asia and the Far East. They will supply the garment factories directly, and this has the benefit of reducing the overall cost of the packaging and ultimately the cost of the product.

A retailer can have many different brands, all with their own unique label and packaging requirements; the role of quality assurance is to help the factory identify what is required for each product. Packaging and labeling should become an integral part of the product specifications, showing all components, and, most importantly, how the finished presentation should look.

There no better method for conveying information than using graphics, where possible, as this not only helps overcome language problems but is easier for operatives to relate to the products that they are working with. Samples of approved packaging can be sent to the factory, and they would make notes and drawings, in case the sample gets misplaced or mysteriously disappears, which often happens. It is infinitely better for you to create your own specification, which you can then circulate to all departments in your company and new suppliers through the Internet by using a standardized format, which the factories will recognize.

I hope the reader will appreciate the importance of specifying every feature of the product and the importance of how we design the format for presenting the information, viewing the information as if we are the end user, and whatever the size of your company, attention to detail is always essential to ensure that we get what we intended.

There are many different ways of packaging, and the manual should include those that your company most often uses. The following items are the most commonly found in the packaging section, and I have shown how they can be illustrated:

- Bags with child warnings
- Bag labels
- Garment's sew-in labels—size fabric composition
- Care label
- Swing tickets
- Different styles of hangers and sizes
- Cardboard inserts
- Methods of folding and displaying products
- Carton specification and sizes (also see Chapter 5)

Each item should have a product reference, allowing your manufacturer to order locally directly from your nominated supplier. When the buyers order new samples, they can refer the factory to the manual for the type of packaging required.

Today, retailers "dress up" their products to show them off and stand out from the crowd, with cleverly designed packaging and attractive swing tickets full of product information. Detailed specifications are needed for the product and accessories, for it to come together as a complete package (Figure 9.17).

- Second page, as shown in Figure 9.18, shows jean display on hanger and labeling.
- Third page, as shown in Figure 9.19, shows jean on hanger.

It is important to specify the size of accessories such as labels and hangers, as it is not unknown for factories to be sent incorrect labels and packaging. This allows the factory to double check they have received the correct accessories before they are sewn or attached to the garments.

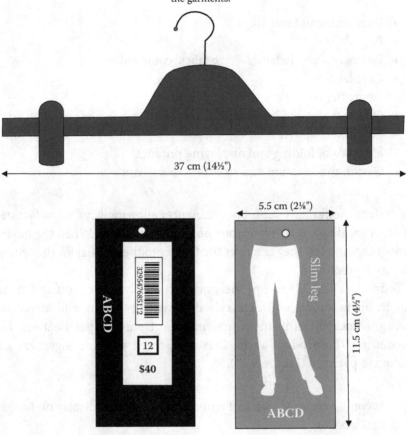

FIGURE 9.17
Trouser hanger and labels.

Supplier's Manual • 129

PAGE 2

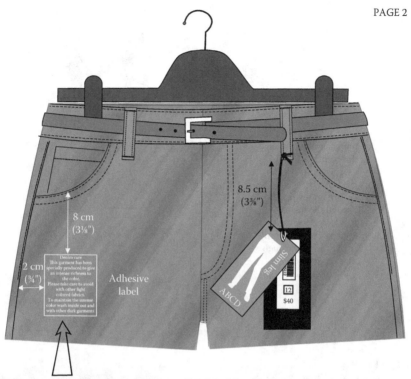

FIGURE 9.18
Jean display on hanger and labeling.

TROUSERS AND JEANS

A good display can help considerably to sell a product. Trousers and jeans can be folded and stored on display shelves, but hanging on rails allows the customer to quickly see the style and handle the fabric. With this style of hanger the front and back of the trouser is fully displayed, as it is this area that the customer is most interested in. The swing tickets identifies the shape of the leg and the price.

The illustration shows how the belt is displayed, the size, text and positioning of the tickets.

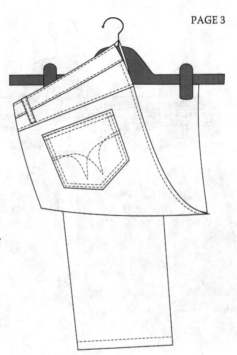

PAGE 3

This is another option for displaying jeans featuring the back pockets. The hanger clips are fastened to the legs and the top of the jean folded back with the curved metal part of the hanger put through the belt loops at the center back.

FIGURE 9.19
Jean on hanger.

TROUSER HANGER WITH BAR

Previously, with the ladies' fashion trouser specification, I showed instructions for folding trousers into a bag and then packed into cartons. Alternatively, trousers can be delivered on hangers, ready to be delivered to the stores. This is a typical instruction to the factory on how to prepare the trouser for delivery, and it is important that the swing tickets are always positioned in the same place on every pair. The main ticket, stating that it does not need ironing and implies "easy care," which is a very good selling point with men's trousers, should be placed almost at eye level so that it catches the customers' attention.

You may want to add further information such as packs to be arranged in size ratios and overbagged with an adhesive label that lists the contents (Figure 9.20).

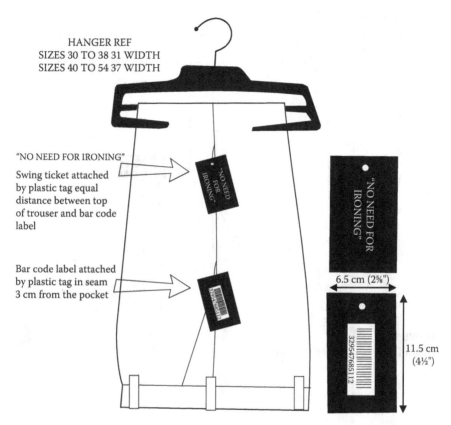

FIGURE 9.20
Trouser on hanger.

TROUSER'S INSIDE

You will probably want to standardize the inside labels' positions for all or most of your trousers, and this format can be used for trousers that have at least three pockets.

It's not always possible, but it does make your life easier if you can standardize your methods of labeling and packaging wherever you can (Figure 9.21).

- First page, as shown in Figure 9.22, shows ladies briefs.
- Second page, as shown in Figure 9.23, shows bag for briefs.
- Third page, as shown in Figure 9.24, shows bag with cardboard inserts.

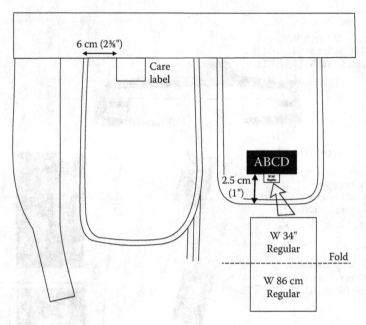

FIGURE 9.21
Inside of trouser with label positions.

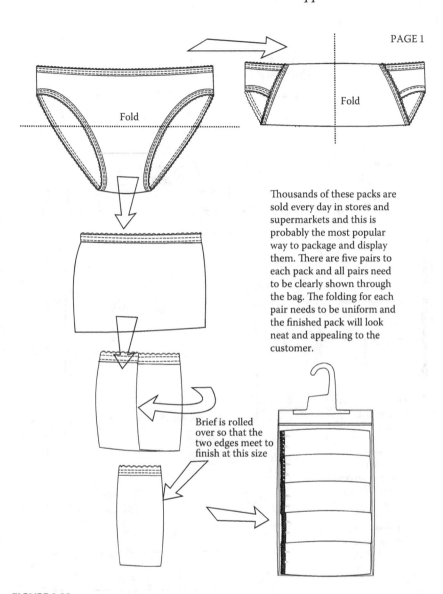

Thousands of these packs are sold every day in stores and supermarkets and this is probably the most popular way to package and display them. There are five pairs to each pack and all pairs need to be clearly shown through the bag. The folding for each pair needs to be uniform and the finished pack will look neat and appealing to the customer.

FIGURE 9.22
Pack of five ladies briefs.

PAGE 2

SAFETY FIRST TO AVOID DANGER OF SUFFOCATION KEEP THIS BAG AWAY FROM CHILDREN AND SMALL BABIES

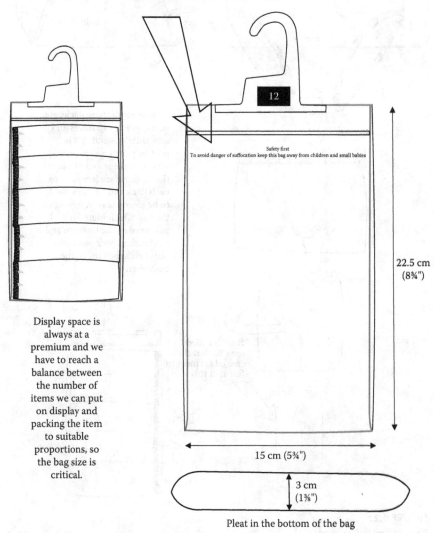

Display space is always at a premium and we have to reach a balance between the number of items we can put on display and packing the item to suitable proportions, so the bag size is critical.

22.5 cm (8¾")

15 cm (5¾")

3 cm (1⅜")

Pleat in the bottom of the bag

FIGURE 9.23
Bag for briefs.

Supplier's Manual • 135

PAGE 3

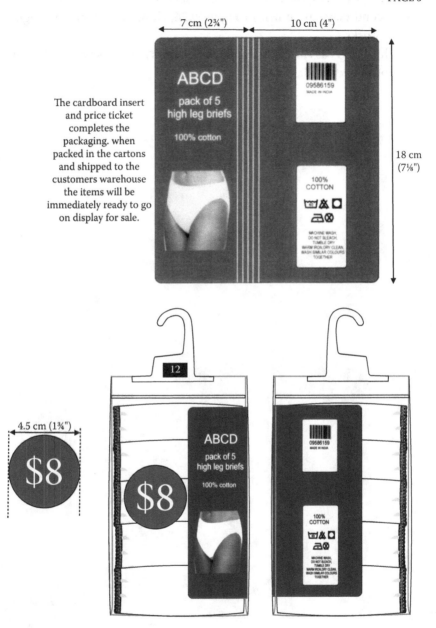

The cardboard insert and price ticket completes the packaging. when packed in the cartons and shipped to the customers warehouse the items will be immediately ready to go on display for sale.

FIGURE 9.24
Bag with cardboard inserts.

- First page, as shown in Figure 9.25, shows pack of five socks.
- Second page, as shown in Figure 9.26, shows cardboard packaging for socks.
- Third page, as shown in Figure 9.27, shows how socks are folded and positioned in packaging.

Supplier's Manual • 137

PAGE 1

The humble sock is sold by the millions every day and is probably one of the cheapest items of clothing that we buy. Over the last few years it has been elevated to a fashion accessory with multi-colored stripes, polka dots, and all types of designs and motifs, and the packaging has had to be re-designed to keep up with the new styling. The new packaging allows the customer to see what is being offered without having to break open the display. In the following specification the factory can see the cut out dimensions of the box, how the socks should be folded and displayed and how the finished product should look. Although all materials will be supplied by the printer, the factory will use the specification to double check that all components are as the customers requirements. This method of packaging is also used for multi-packs of underwear and pyjama tops and bottoms.

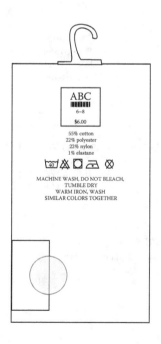

FIGURE 9.25
Pack of five socks.

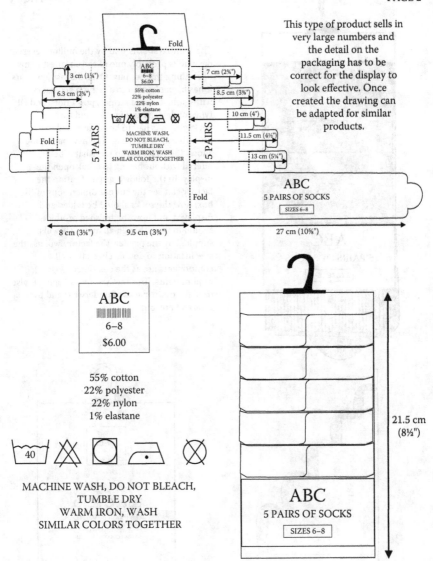

FIGURE 9.26
Cardboard packaging for socks.

PAGE 3

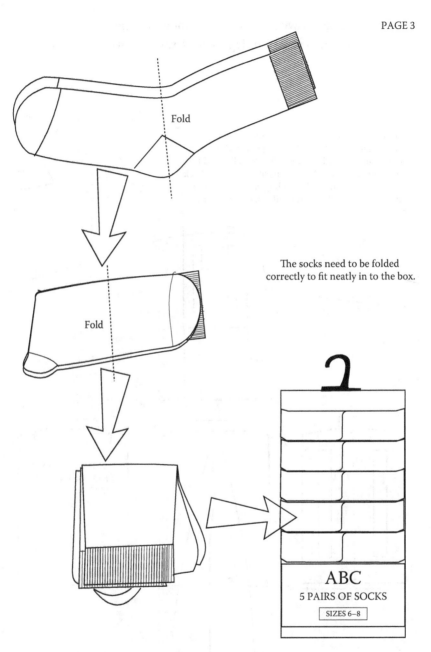

FIGURE 9.27
How socks are folded and positioned in packaging.

- First page, as shown in Figure 9.28, shows folding shirts.
- Second page, as shown in Figure 9.29, shows shirts folded into a bag.

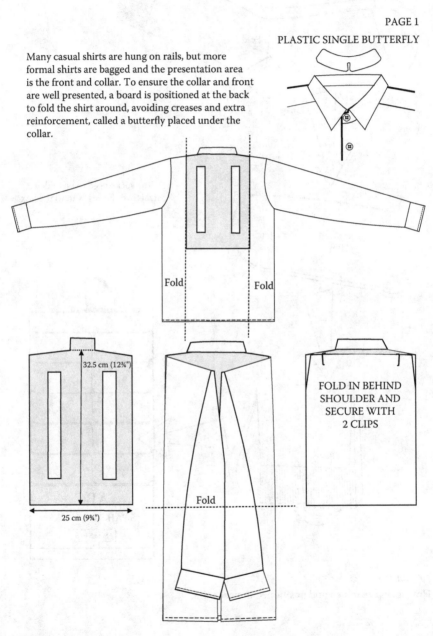

FIGURE 9.28
Shirts packaging.

PAGE 2

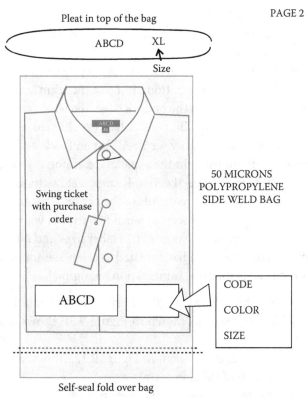

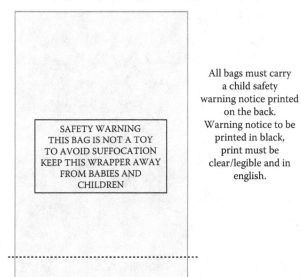

FIGURE 9.29
Shirts folded into a bag.

The next illustrations are specifications for a gilet, which belongs to the family of casual coats and jackets and the following could apply to many different styles.

If we specify that the care label is sewn into the side seam, left side as worn, it could be sewn at the bottom or top of the seam and anywhere in between. If it's sewn 6 cm from the bottom edge, the label can be read without undoing the garment. Equally, the size tab is positioned, so that the customer and the warehouse can easily see it. All brand labels and swing tickets must be positioned uniformly in the same place; if not, it gives a bad impression to the customer. The specification is completed with folding instructions and carton details. Even the details of the carton label are important to help the delivery be quickly processed when it arrives at your warehouse.

Wherever possible, standardize label sizes and positions, so you can use the specifications you have created as templates for other styles and all that is required is the label information changing.

- Insert first page for gilet labeling as shown in Figure 9.30.
- Second page, as shown in Figure 9.31, shows label positions at neck for gilet.
- Third page, as shown in Figure 9.32, shows care label position at the bottom of the side seam.
- Fourth page, as shown in Figure 9.33, shows folding and bag instructions for the gilet.

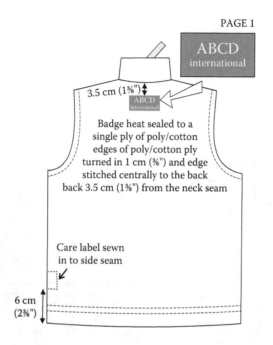

FIGURE 9.30
Gilet labeling.

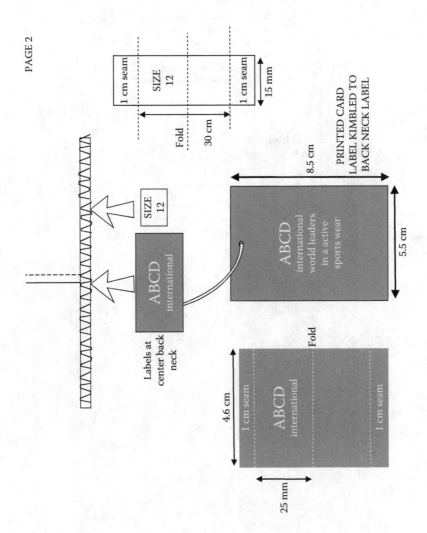

FIGURE 9.31
Label positions at neck for gilet.

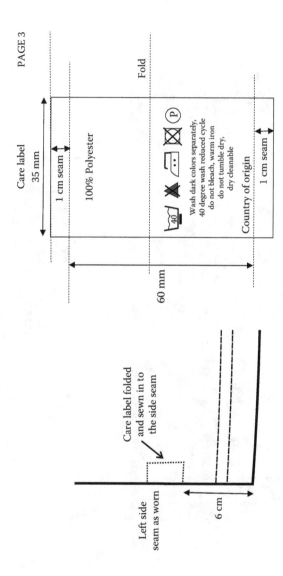

FIGURE 9.32
Care label position at the bottom of the side seam.

146 • The Fundamentals of Quality Assurance in the Textile Industry

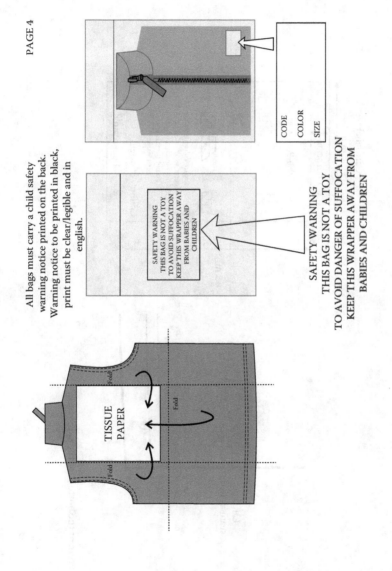

FIGURE 9.33
Folding and bag instructions for the gilet.

STANDARD CARTON SPECIFICATION

- Flat packed garments
    - Merchandise must be packed in individual bags, with bar code and price clearly visible.
- Carton specification 150 gsm semi chem "B" flute double wall (5 ply).
    - Maximum weight 12.7 kg (30 lb). Never overpack even if maximum weight allows.
    - Only one reference number and one option may be packed in each carton; never mix options.
    - To prevent damage when opening with a sharp knife, a cardboard insert must be placed on top of the contents before sealing.
    - Never use staples to close carton; tape must be used on the top and bottom of the carton.
    - Never lose fill to take up room inside the carton.
- Do not use metal straps.
    - Each box must have an authorized box end label; see example with information required (Figure 9.34).

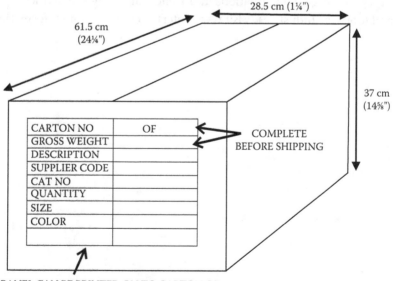

END PANEL CAN BE PRINTED ON TO CARTON OR
LABEL GLUED SECURELY WITH FULL INFORMATION

FIGURE 9.34
Standard carton specification.

ACKNOWLEDGMENTS

My thanks to Intertek, an internationally recognized testing house, that has verified all fabric performance standards referred to in this book.

Intertek supports UK and international fabric, textiles, apparel, and footwear retailers; brands and manufacturers respond to ever-changing governmental regulations and increasing consumer demand by providing fast and efficient performance and quality assurance, testing, inspection, and certification services to ensure that their products meet safety, regulatory, quality and performance standards.

These services minimize the risk and protect the interests of both retailers and consumers, while enhancing brand loyalty and ensuring consumer confidence.

Intertek's state-of-the-art equipment in the UK's largest textile testing laboratory in Leigh and throughout Intertek's global textile and footwear laboratory network enable Intertek to offer a full range of physical, chemical, and regulatory testing; inspection; and auditing services to the industry.

In addition to testing, Intertek's experts offer unparalleled knowledge and expertise on global regulations and trends, affecting the textiles, apparel, and footwear industries. Visit www.intertek.com for more information.

10

Product Development

Product development is an important aspect of quality assurance and is a continuous process, extending to a company's products by range and individual garments.

Companies cannot afford to stand still and ignore new trends in styles and fabrics; buyers will always be looking at what is new in the market and what their competitors are selling.

The quality assurance or technical department would be involved at an early stage in this process, documenting and assessing the developments following through to the finished new product.

Create a file immediately to start a history of the product, saving all correspondence as well as all technical information. Work to a critical path, keeping everyone who needs to know updated.

There are no easy shortcuts to developing new products, and it is better to take small steps quickly rather than trying and rushing completion in one big step.

REVIEW OF A JEANS RANGE

For many seasons, a company might be happy with the range of denim jeans they sell, but then they might realize that it needs updating, because the leg shapes that they offer might no longer be fashionable and not what the customer might be expecting. Another reason for a review is that as the number of jean styles your company offered have grown, they might have been sourced from different suppliers, and there was no consistency in the fit, as each supplier worked on their own pattern block, realizing that although the leg shapes are different, the basic fit should be the same.

This example shows a development of a more fashionable range of ladies' jeans and the three leg shapes chosen for the core range. They might decide to buy in jeans from a competitor, which are representative of the fit required. The first step is to create a size chart, with measuring points incorporating the three styles, and show how the leg shapes are to be interpreted. The second step is to redesign the front and back style of the jean by adding the pocket shapes, sizes, and different styles of decorative stitching for the back pockets. Back yoke and belt loop details should also be added. This information can now be sent to a factory to make the initial samples. Only when entirely happy with the fit of the sample should a full size chart be graded (Figure 10.1).

- Second page, as shown in Figure 10.2, shows the size chart.
- Third page, as shown in Figure 10.3, shows the measuring points.
- Fourth page, as shown in Figure 10.4, shows front pocket, zip, and belt loop measurements.
- Fifth page, as shown in Figure 10.5, shows the back of jean with hip pockets of different styles.

Product Development • 151

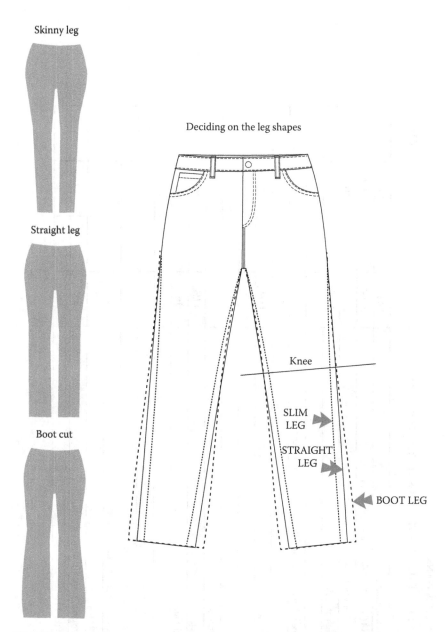

FIGURE 10.1
Deciding on leg shapes.

ABC LADIES JEANS

DEVELOPMENT SIZE CHART   UK SIZE 12 PETITE FITTING   PAGE 2

	Inch	Cm
A. Waist opening	32"	81.3
B. Hip 10 cm (4") down from waist band seam	37½"	95.2
C. Hip 20 cm (8") down from waist band seam	39½"	100
D. Thigh at fork	22½"	57
E. Knee midway + 7.5 cm (3") up toward the fork		
Slim leg	15½"	39.5
Straight leg	17½"	44.5
Boot cut	17½"	44.5
F. Bottoms		
Slim leg	14½"	36.3
Straight leg	**16"**	40
Boot cut	19½"	49.5
G. Front rise from top of band	11"	28
H. Back rise from top of the band	14"	35.5
I. Inside leg	28"	71
J. Side seam from top of band	39"	99
W/B DEPTH	1⅝"	4

FIGURE 10.2
Size chart.

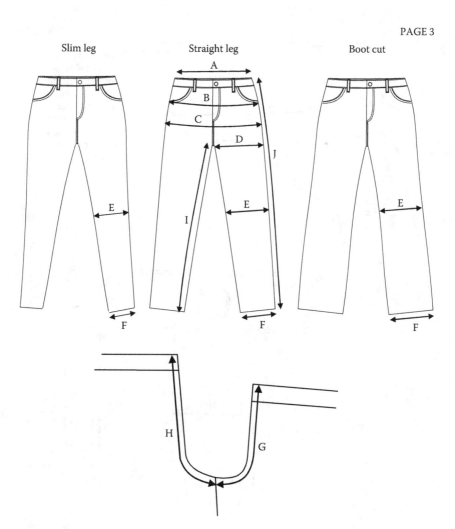

FIGURE 10.3
Measuring points.

PAGE 4

Decide on pocket shapes and sizes, belt loop positions, fly stitching, back pocket sizes and decorative stitching–all these points make a garment aesthetically more pleasing

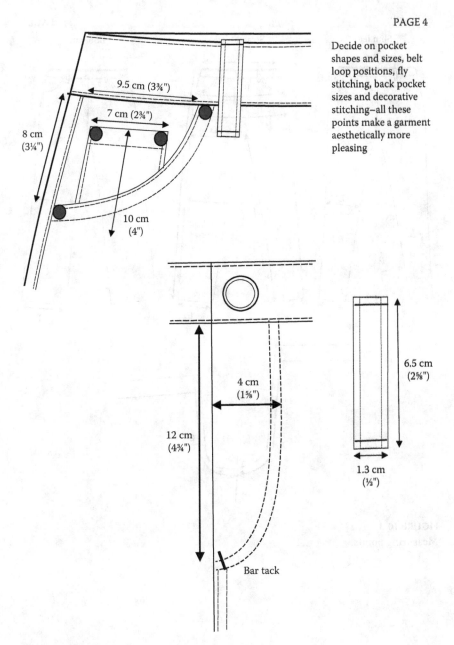

FIGURE 10.4
Front pocket, zip and belt loop measurements.

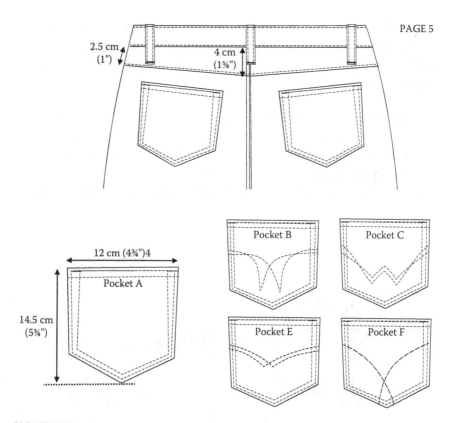

FIGURE 10.5
Back of jean with different style hip pockets.

GIRLS' DRESS DEVELOPMENT

Starting or developing a children's range has certain difficulties, particularly with sizing. In this instance, I started with two bought-in samples of the dress for ages 5 and 12 years, in the smallest and biggest sizes, that were to be offered in this style. The next step was to try them on girls of these ages and decide if you are happy with the fit, note any changes required, and then create a size chart from these two sizes. When the factory samples are approved, the full size chart can be graded. With adult clothes, there is usually someone in the office or another department that we can call on at short notice to try on a sample, but children's fit sessions need to be arranged in advance to fit in with parents' busy schedule, and as children grow, we need to keep recruiting new models.

As well as measurements and style consider the packaging and how you visualize the dress looking on display, build in as much detail as possible to send to the factories, so they can work out accurate prices (Figure 10.6).

- Second page, as shown in Figure 10.7, shows construction details.
- Third page as shown in Figure 10.8, shows labels and hangers.
- Fourth page, as shown in Figure 10.9, shows measuring points.
- Fifth page, as shown in Figure 10.10, shows the size chart.
- Sixth page, as shown in Figure 10.11, shows dress in hanging bag.

FIGURE 10.6
Girls' dress development.

PAGE 2

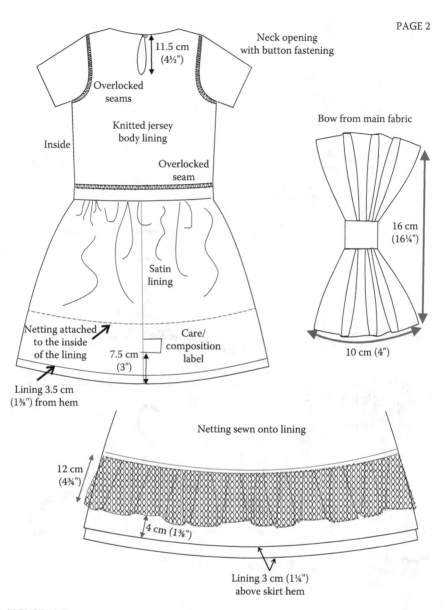

FIGURE 10.7
Construction details.

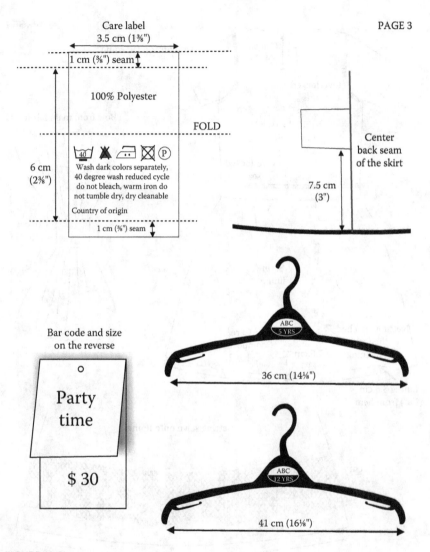

FIGURE 10.8
Labels and hangers.

PAGE 4

FIGURE 10.9
Measuring points.

ABC GIRLS' PARTY DRESS
DEVELOPMENT SIZE CHART

PAGE 5

	5 YRS			12 YRS	
	CM	INCH		CM	INCH
A. Chest flat measurement	31	12¼"		40	15¾"
B. Waist flat measurement	**30**	**11⅞"**		**39**	**15⅜"**
C. Hem sweep flat measurement	57.5	22⅝"		75	29½"
D. Neck opening	13.5	5⅜"		17	6⅝"
E. Shoulder length	4.5	1¾"		6.5	2⅝"
F. Sleeve opening half measurement	10	4"		13	5⅛"
G. Neck point to waist seam	25	9⅞"		35	13¾"
H. Armhole half measurement	15	6"		20	7⅞"
I. Waist to hem	32	12⅝"		39	15⅜"
J. Overarm	13	5⅛"		17	6⅝"
K. Back neck drop	2.5	1"		2.5	1"
L. Front neck drop	6	2⅜"		8	3⅛"

FIGURE 10.10
Size chart.

Product Development • 161

An example of how you want the product packaged could be sent to the factory at this stage.   PAGE 6

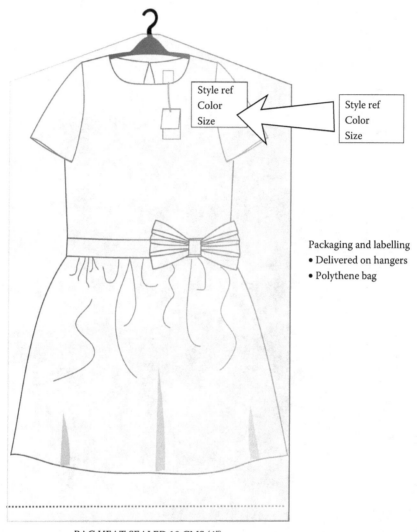

FIGURE 10.11
Dress in hanging bag.

11
Managing the Critical Path

Having established the physical properties of the garment, the next procedure is to chart how we are going to ensure that we get what we specified within the required timescale. The critical path charts the progress from initial sample to finished delivery, but the most important date is when the delivery is required. Below I have set out a typical timetable, working backward from the delivery date. Each step is essential and reflects a realistic time allowed for each procedure.

- *Delivery to the stores*: This is the most important date in the critical path; careful planning will go in to the date for delivery to maximize sales. It could be a launch for the new season or a new range to attract customers into the store midway through a season. Missing the window because of late or faulty deliveries could have a significant impact on sales and profitability; companies could impose penalties or even cancel orders if not delivered on time. It would make sense to be on the safe side and arrange to have deliveries slightly earlier.
- *Delivery to your warehouse*: A week should be allowed for inspection and any repackaging that might be required before delivery to the stores. Time should be allowed for a 100% inspection of the delivery and rectification, if necessary. It's better to deliver part of the delivery correct than the whole of the delivery with a high fault rate.
- *Shipping time*: This can vary, depending on where the goods are coming from and how they are to be shipped: by boat, flown in, or part by sea and part by air. On average, we would allow 2–3 weeks. This should also include any delays at customs.
- *Production samples*: Production samples should be received approximately 1 week before delivery to be shipped. These samples will act as an early warning if there is a problem with the bulk of the

production. If these samples are completely wrong, then the delivery can be cancelled before payment is finalized. However, depending on the nature of the problem, there could be time for the factory to do 100% check and rectify faults before the delivery is shipped.
- *Manufacturing time*: This depends on the size of the order and the number of machinists on the production line who are making your garments; the factory will advise the time to be allowed, but on average, this could be approximately 2 weeks.
- *Delivery of material to the factory*: Production material should be ordered in time to arrive well before production starts, allowing time for the factory to examine the fabric before cutting and send a sample of bulk fabric for testing and, when requested, send sample fabric to the customer for approval. The fabric may be held in stock by the factory or a local supplier, but if it is being made especially for your order, you need to keep checking its progress with the factory.
- *Sealing samples/initial samples*: Initial samples are defined as the first prototype made by the factory, which might be in a substitute fabric and may need amending. Sealing samples are normally the stage after the initial samples, with everything corrected and in the right fabric. Initial samples or sealing samples allow approximately 2–3 weeks, particularly if the factory is developing the patterns and sourcing trims. Allowing sufficient time at this stage can help avoid delays later.

In this example, we are allowing approximately 10 weeks from the start to finish. This is only a guide, as the timetable indicated depends on the type of garments and will vary from factory to factory and from country to country. The buying department that places the contracts and the manufacturer should know that when a schedule is agreed, it has to include provision for sets of samples and fabric to be approved before production can start and that these procedures cannot be bypassed.

The critical path form is for multiple products, all from the same factory. At a glance, you can see the progress that this factory is making. This is a powerful tool to be used in quality assurance, ensuring that each step is completed in the correct timescale.

Responsibilities for chasing suppliers for samples and updates vary from company to company. In some, it can be the buying department, whereas in others, it can be the quality control department. In whichever way your company works, QC and buying should have regular meetings to discuss progress and any potential problems that are causing delays (Figure 11.1).

CRITICAL PATH FACTORY A

	SEALING SAMPLES REQUESTED MAY 23	SEALING SAMPLES DUE JUNE 9		BULK FABRIC DUE		PRODUCTION DUE TO START STARTED JUNE 24	PRODUCTION SAMPLES DUE JULY 1		DELIVERY DUE TO BE SHIPPED ON JULY 8	DELIVERY TO THE WAREHOUSE ON JULY 25	DELIVERY TO CUSTOMER OR STORES
		RECEIVED	APPROVED	RECEIVED	APPROVED	DATE STARTED	RECEIVED	APPROVED	DATE SHIPED	DATE RECEIVED	
STYLE 1	23-May										01-Aug
STYLE 2											
STYLE 3											
STYLE 4											
STYLE 5											
STYLE 6											
STYLE 7											
STYLE 8											
STYLE 9											
STYLE 10											

COMMENTS

FIGURE 11.1
Critical path.

- How to use the critical path effectively?
  - The form should be e-mailed to the factory to complete.
  - The factory should be contacted for updates every day if the information is missing.
  - A copy should be circulated to the buying department or available for it to see every day.

I cannot overemphasize the importance of keeping to the agreed timetable when samples should be received. Planning production is always problematic for a factory, as there are many factors that have to be taken into the account: First, production cannot be given a start date until they know for certain when all fabric and trims will be in the factory, and second, if their customers have not approved samples. However, it is not unknown for factories to start production before sealing samples are approved, because they are short of work or getting behind with their orders, and this is a textbook example of how things can go badly wrong and quality is the first casualty.

If there is a delay in sending information or samples, investigate why. Keep digging until you get a satisfactory answer. Delays could be hiding a problem that the factory is reluctant to tell you about.

Today, we can always keep in daily contact with our suppliers by e-mail, Skype, land line, or mobile, even outside office hours. All manufacturers are hungry for orders and want to see their companies grow, especially those in developing countries, many of which are still family run. The owners are often available to talk round the clock, and I have worked for companies that have contacted their suppliers during the night if they have a problem that is bothering them. Personally, I think that's going too far, but it shows the level of commitment some people are prepared to show to ensure that factories meet their deadlines.

It is not sufficient to write procedures for quality assurance; they will not work on their own. Like while driving a car, you have to be at the wheel, guiding it to its destination, and you have to be prepared to go the extra mile to make sure that you arrive where you are intended to be.

Discussions with your suppliers about quality should always be positive and should define quality with strengthening and increasing business, and this should be a learning curve for both of you. If you expect your suppliers to react quickly in producing samples, you have to reciprocate and equally respond quickly to their queries and concerns. When they send in samples for approval, you have to set a time limit for yourself to check the samples and reply back to them. I would suggest sending a report back to the factory on samples within 24 hours of receiving them. This is crucial, as any delays on reporting back to the factory can very easily have a knock on effect and very quickly cause delay in deliveries. If you do not respond quickly, you will be blamed for the eventual delay, so do not give the factory a reason to point the finger at you, as you have to set the example of quick response.

12

Sample Reports and Approval

Before production can start, sealing samples must be approved for style, size, construction, and quality of workmanship. One sample must be returned to the factory with a sealing tag attached and a sealed sample should be kept in the buying office. The seal identifies the sample that was approved and ensures that the factory cannot substitute it with another sample. These samples are an essential procedure, as now, the factory and the buyer have an identical point of reference, particularly if there are any quality issues, and the factory can now start production. First, it is the factory's responsibility to make sure that samples are correct, and then, it is the responsibility of the technical designer/quality controller to ensure that the samples meet the accepted standard required by the buying company. Of course, no sample is ever perfect, and I always find something to comment on! Minor faults can be accepted on sealing samples if they are backed up with the comment "please ensure this is corrected for production" and you make a point of reminding the factory about ensuring that the production will be corrected. However, where do we draw the line? Moreover, this is an issue that I refer to at different times in this book, as this particular subject does need discussing in detail. Rejecting samples is a sensitive issue, because rejection immediately relates to delay of production. It does not have to be considered a sensitive issue, but that's how it will be perceived. If factories have to remake samples, it is not a disaster; they can make up the time later. The decision to reject has to be carefully considered, and if you believe that accepting the samples with serious faults or mistakes, if not corrected, could risk the delivery being rejected, then it is the right decision. It is a good practice to talk to the factory first and ask if they were aware of the problem when they checked the samples, and if they were aware of the problem, why did they not mention it in the report or, even better, remake the sample before sending. If the factory were unaware of the faults, *why not!* This would be the situation that would cause the most concern that a factory cannot follow an instruction or recognize a major fault. You need to then describe the problem in detail and send a report on how it must be rectified.

Often the critical path does not run smoothly, and when this happens, it is essential that you are seen to be taking the right course of action, because rejection costs time and money, but it is the correct course of action if the weight of probability indicates that bulk production would be faulty, which in turn will cost the buyer and supplier much more.

When suppliers send samples for approval, you should insist that they send a sample report filled in by the factory's QC. I have come across many instances when a factory makes samples and then rush them off to the customer without checking them against the specifications, and when you inspect them, a major fault is found; this is a total waste of valuable time. If they had checked the samples first, they could rectify or remake saving at least a week. The other advantage is that you can check if they are measuring the garments correctly and if the factory is also checking the other points listed on the inspection form.

Figure 12.1 is an example of a form that I personally prefer, as it can be used for both samples and deliveries.

Apart from the chart to record measurements, there are six boxes that cover separately the different aspects of the garment. All boxes should be completed with comments, where necessary, and ticked if acceptable or not, and you should insist that the factory should use the same form and fill in every section, which will help them focus on each area of the garment.

Often, we need to inform the factory about faults or amendments to the samples, and the text is not sufficient. Through webcam, we can show faults in garments, but it is also a good idea to support your comments with a visual record. I have shown two examples here.

The rib on the sweatshirt (Figure 12.2) was not acceptable. I took a photo and transferred it to a CAD package, enabling me to draw and add text to the photo.

Figure 12.3, where it was decided after fitting the first sample that the pattern for the trouser needed amending. This could be explained in the text, but to reinforce this with a diagram is easier for the factory to understand.

Figure 12.3 shows amendments to a trouser.

SAMPLE/DELIVERY REPORT

REF No

DOCKET No

DESCRIPTION

FABRIC/COLOR ACCEPTABLE ☐ NOT ACCEPTABLE ☐	STYLE ACCEPTABLE ☐ NOT ACCEPTABLE ☐	COMPONENTS ACCEPTABLE ☐ NOT ACCEPTABLE ☐
CONSTRUCTION ACCEPTABLE ☐ NOT ACCEPTABLE ☐	WORKMANSHIP ACCEPTABLE ☐ NOT ACCEPTABLE ☐	LABELLING/PACKAGING ACCEPTABLE ☐ NOT ACCEPTABLE ☐

SIZE												
	CHART	SAMPLE	CHART	SAMPLE	CHART	SAMPLE	CHART	SAMPLE	CHART	SAMPLE	CHART	SAMPLE

COMMENTS

QUANTITY IN DELIVERY

QUANTITY INSPECTED

SAMPLE/DELIVERY
ACCEPT ☐
REJECT ☐

ACTION TO BE TAKEN

SIGNED

DATE

FIGURE 12.1
Sample/delivery report.

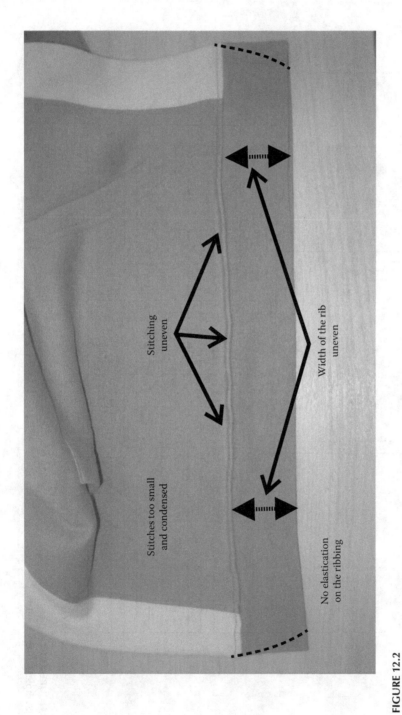

FIGURE 12.2
Amendments to sweatshirt.

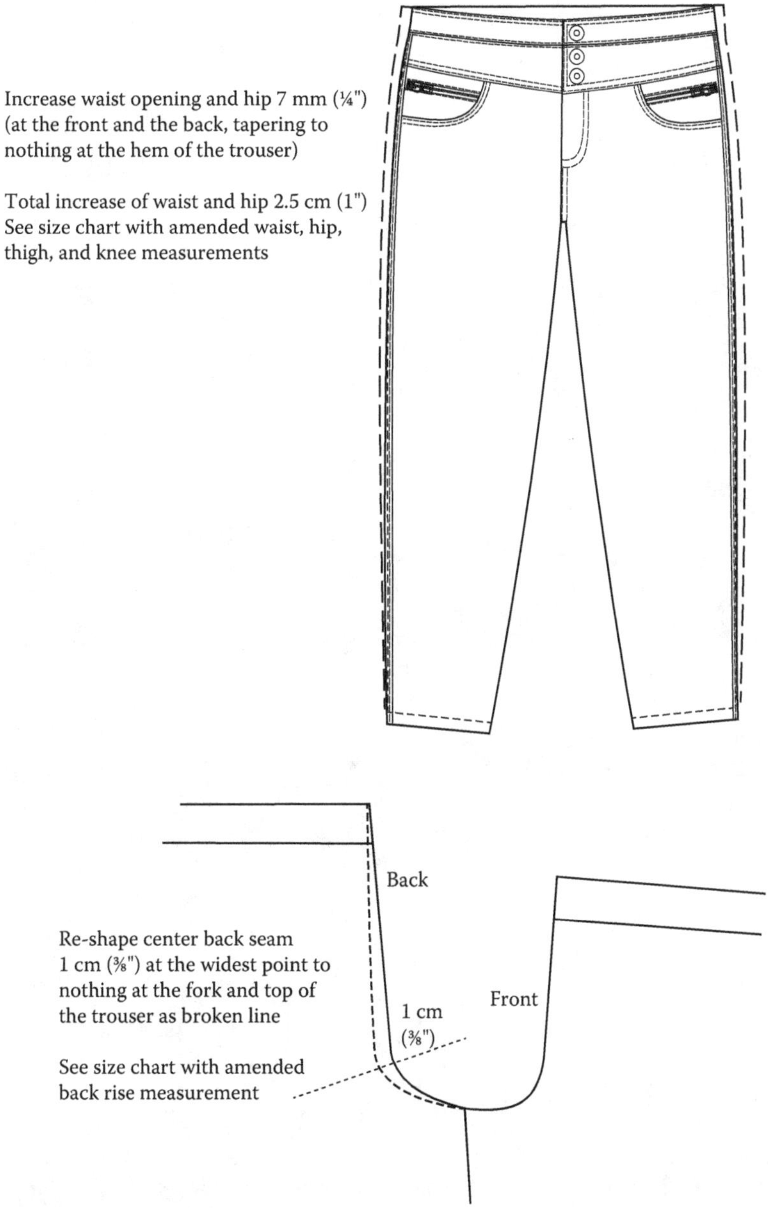

Increase waist opening and hip 7 mm (¼")
(at the front and the back, tapering to
nothing at the hem of the trouser)

Total increase of waist and hip 2.5 cm (1")
See size chart with amended waist, hip,
thigh, and knee measurements

Back

Re-shape center back seam
1 cm (⅜") at the widest point to
nothing at the fork and top of
the trouser as broken line

1 cm (⅜")

Front

See size chart with amended
back rise measurement

FIGURE 12.3
Amendment to a trouser.

13

Assessing and Working with Factories

We have to work hard to build a good relationship between our company and our suppliers, based on cooperation and trust, and to be realistic, the amount of cooperation we receive is linked to the amount of the factory's production capacity we take. If we are in the position to take approximately 30% or more of their capacity, the factory will be willing to cooperate on matters of quality, because financially, your business is valuable to them. Other times, we might have to spread our orders thinly over many suppliers and they will judge you against their least demanding customers, who may not ask for as many samples or insist on as much fabric testing as you require. Factories often work with a mix of customers, each with a different set of priorities and standards. This may cause difficulties at first, so you need to explain who your customers are and the level of quality that they expect to ensure that they keep coming back to buy your products, which in turn will enable you to place more business with their company. If this sounds like the carrot and the stick, it is meant to! If you want a factory to view quality as importantly as you do, then motivate them in a positive way, patiently explaining that if they work together with you to get things right, there's every possibility that business will increase. Always take the opportunity to talk about your company in a positive way and consider your suppliers as an extension of your business.

Keep an up-to-date history and have a file with every e-mail, specification, sample report, and piece of information about each product, because you will not be able to remember what was agreed 2 months ago by whom and when.

In many instances, a quality controller will be expected to visit the overseas factories they are working with. This might be as and when it is thought necessary (there might be quality problems or there is an unusually large amount of production going through the factories at one time) or the visits might be arranged on a regular basis to coincide with final approval of sealing samples or midway through production.

Ideally, approving samples at the factories is preferable for many reasons; the factory sees firsthand how you measure and check a garment. Problems can be resolved on the spot with the pattern cutter and sewing room management. It might be possible for samples to be remade while you are there, and time wise, you can be saving several weeks as well as having a much greater chance of preventing the problems from reoccurring. There is, of course, one drawback to doing this, and that is, if necessary, being able to try the samples on a body. Sometimes, fortunately, the technologist might be of the correct size and body shape. Other option I know companies have used is to take with them to the factories the person they normally try garments on or hire a model locally in the country of manufacture.

The first priority is the line of communication between yourself and the factory. The factory should designate a member of the management, who will act as your opposite number in the factory and will be responsible for all quality issues and who you can talk to on a daily basis. Your first visit to the factory will be the perfect opportunity to get to know that individual and familiarize yourself with the factory. You should also understand that often factory management in third-world companies have not visited Europe or America and are unaware of the general quality of apparel for sale in these countries. Their experience might only be of the home market, which is generally of a lower quality. So, although they might not say it outright, they may think that your quality standards are high. Earlier in Chapter 2, I wrote the following: "The challenge is, working with factories to ensure sure they see quality as if through your eyes and make the decisions you would make even if you are thousands of miles away and if they are unsure of the right course of action contact you immediately." I mention this again because this applies to everyone in the factory, and if your normal contact in the factory is away, whoever takes over at that time should be singing from the same hymnbook.

You may visit overseas factories with a preconceived idea of what they should be like; often, you will not be disappointed, but sometimes, they might not be what you expected, so you need to keep an open mind. The important thing is how they function, and at the end of the day, do they deliver the goods. Areas to assess in the factory are as follows:

- Fabric storage and inspection
- Pattern-cutting facilities
- Cutting room and equipment
- Sewing room
- Pressing
- Final inspection
- Storage and packing

You are looking primarily for "good housekeeping," cleanliness, a good workflow, well-lit work areas, and effective quality control (QC) through all operations to final inspection. Most importantly, look at the work going through the factory, as this is a snapshot of their quality and the customers they are making for. A supplier might tell you that he make for a well-known brand and show you a sample with that company's label, hanging in the showroom, but the only guarantee that he does make for that brand is if he is manufacturing for that brand at that moment.

The buyers in many of the companies I worked for did not often visit factories but initially made contact with manufacturers by visiting major trade shows. This is an opportunity for them to see existing suppliers and source new ones, all under the same roof, saving many hours of traveling time. They would trial new factories with a small order, and if that were successful, they would place further business.

REPORT FROM CHINA

Stanley Bernard Brahams

The following is a report that I wrote on a trip to China when visiting existing and new suppliers. On this visit, I was working for an import agent, with offices in Manchester and London (see C in my company profile section) The company had not previously sent quality controllers to the factories and all communication was by e-mail and phone, but as business increased, the company realized the importance of quality assurance having direct contact with their suppliers, and both offices used the same factories. Our customers were a mix of mail order companies and stores and the merchandise was mainly young fashion, and I was responsible for the quality of the merchandise bought through the Manchester office. Armed only with a list of the London office merchandise, my brief was to report back on my impressions of the factories and any potential problems to the London QC department.

Both head offices would want updates every day and each night when I got back to the hotel, I would report back to the UK and then complete a written report to be circulated at the end of the trip. A copy of the report would be sent to each factory on their merchandise. It is highly recommended that a quality controller visits factories as often as possible, as many issues and concerns about procedures and quality are raised when talking face to face, which might not surface when talking on the phone or e-mailing from your head office.

For the quality controller, it means many hours traveling from factory to factory. So, you have to plan your itinerary very carefully to make sure that you have time to visit everyone you need to see.

It is best to keep the comments short but to the point and be sure to report back on any concerns that you may have that affect the quality of the merchandise and any potential difficulties you may find while trading with any particular supplier.

During the trip, I took the opportunity, when possible, to visit their fabric suppliers to learn more about how the fabrics were made and to discuss quality issues with the mills. When visiting denim manufacturers, a percentage of the production would be at the laundry for washing, stone washing, and sand blasting. So, a visit to the laundry was always a good opportunity to learn firsthand about these processes and check if they were achieving the results we required. On one such visit to a laundry, I was shown the sand blasting operation. It was amazing and I had no idea that this was how it was done! (See Figure 13.1.)

Assessing and Working with Factories • 179

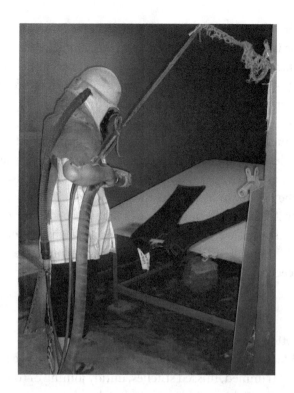

FIGURE 13.1
Sand blasting.

FACTORY A

- Three styles in production for mail order.
- Fabric on one style shrinking more than allowed requested for 100% inspection, and factory was able to press smaller garments to chart measurement without fabric relaxing.
- Factory was genuinely concerned that the number of samples requested sometimes before final approval may cause delay in shipping, causing the delivery to be cancelled.
- On occasions, they had sent shipment samples for approval before ordering the bulk fabric, in case the order is cancelled.
- I made the point strongly to the factory that they must follow the critical paths requested by head office. If they felt there might be a

problem, then they should contact head office immediately and discuss the issue. They have agreed to do this, and I believe that speaking to the factory on the phone if QC has any concerns helps.
- Factory A has a program of 12 new styles for mail order and retail customers, where samples are required urgently.
- Sometimes, they take the view that samples are a nuisance. I explained that every sample is a potential order; delay in making means delaying approval and that can affect the order being placed. If they have many samples to produce, they should use machinists from the factory to help make them.
- Factory A has agreed to pack goods that we specify for mail order in cartons, so that we do not have to repack when the goods arrive in the UK. There are a few points left to finalize before he will be ready to proceed.

ITR 318

- Production being packed.
- Asked factory to reinspect because of the following reasons: cotton ends not trimmed, missed stitches, untidy joining of rows of stitches, and untidy stitching at front waist band.
- Jeans were not being packed to customers' instructions.
- Factory was still using fake YKK zips. I told them that they had to use genuine YKK zips or zips that have been tested by a recognized lab and that fake zips were never acceptable.

ISK 326 Black Mini with Diamante's

- Preproduction washed blue instead of black; color will be corrected for production.
- Measurements within tolerance.
- Back pockets not level with yoke seam and drawing sent not clear; please confirm to factory the position of pockets.
- Samples sent to head office on Wednesday.

ITR 318 for Bigger Sizes

- Factory says that the order of 300 pieces is not big enough to order minimum amount of fabric, but they have ordered fabric and are expecting orders to use up extra. Has size 20 been approved?

- Fabric ready on 20th July—ship the order at the end of July. Merchandiser sent critical path for all styles.
- They are moving to a new factory, and at present, they have approximately 100 machinists. They intend to increase this to 200 when the dormitories for their workers are finished. This will increase production from 25,000 per month to 50,000 per month.

FACTORY B

- Unable to see production ITR 254, as factory is in the South China.
- I visited two local factories, where they are involved in a joint business venture and are making for Europe and America.
- First factory was a woven cloth factory—This company produced high-quality up-market casual and evening wear, with a lot of design content. Fabrics included silk, chiffon and georgette, and linen and cotton blends.
- Second factory was a technically very good knitwear factory with computerized flatbed machines and hand-knitting machines, able to produce fully fashioned and a variety of fine-knit wool blends, acetate and acrylic.
- Specifications have been sent to them to quote prices for our customers.

FACTORY C

- Checked IJK171 hooded jkt.
  Majority of jkts were approximately 2.5 cm short in the length. Insufficient allowance had been made for the body ruching.
- Color white had an off-white grey wadding, which showed through the outer fabric, spoiling the appearance of the garment.
- My overall impressions were that this was a good factory with approximately 200 machinists. General sewing quality was good, producing wovens, mainly jkts of various styles and fabrics. These were made mainly for Europe. I brought back the following samples:

Mock suede gilet.

Satin cotton tailored ladies' jkt lined with shoulder pads. This may be of interest to a chain store who have asked us to source ladies' soft tailoring.

FACTORY D

Checked jkt IJK163.

- Garments corresponded to the specification.
- Would recommend with woven garments to specify a weight of outer fabric in grams per square meter and wadding in grams or ounces, not by garment weight, especially as the outer fabric of this garment was very fine.
- Main factory's 500 machinists will be increasing to 1000, making mainly for Europe.
- Strong on design and development, making all types of quilted and padded jkts, including imitation suede.
- Ruched panel jkt (see photo).

Our order was subcontracted to another smaller factory.

FACTORY E

- Main factory's 180 machinists and smaller unit's 90 machinists.
- Good-quality workmanship, making mainly for Europe.
- Making a variety of jkt styles in different fabrics.
- Cotton padded and down jkts. Lined denim, also skirts, and imitation suede.
- Cotton skirts with zips (brought sample back).
- Enzyme wash cord jkts with fur collar and cuffs (see photo).
- Long cord coat (see photo).
- Strong on design and development.

FACTORY F

- Export company with several factories in the group.
- Supplies leather products.
- Supplies knitwear.
- Supplies padded jkts.
- Supplies linen mix trousers and jackets.
- Good styling and design content.
- Supplies European and American brands such as Red Skin, Big Star, and Basla & Lucia.

- Factory sending photos of various styles.
- Visited factory producing padded jkts.
- 350 machinists; the factory had own embroidery machines.
- (See photos of jkts with circular quilting.)
- Good-quality workmanship; new very clean factory.

FACTORY G

- Main factory in Shanghai, 150 machinists, good standard of work, producing mainly denims but also some knits (I brought back t-shirt sample with print).
- This factory does a lot of design and development and uses different trims and accessories on the garments.
- I brought back a suede-look skirt with stud design and denim jkt with lace and diamante trim.
- Factory works with a variety of fabrics, including stretch denim, rain denim, and cross hatch.
- Production going through for UK mail order and various fashion retailers.

FACTORY H

- Samples of KN 102 were brought to the hotel; we did not have specification.
- Maggie tried samples on; fit was okay.
- Knitting and workmanship okay; labels were not attached.
- Production finished shipping on 15th by air.
- 1000 black and 500 each of white and pink sent, one of each color, to the London office on Wednesday.
- Factory sending samples of KN 110 to the London office.

FACTORY I

- Beaded ponchos for stores.
- Garments were being packed, docket 4173, shipping 13 July, individually packed in a bag and four in a larger outer bag.
- Weight 280–300 g.
- Measure to chart; knitting even and no faults; sequins secure.

- Swing ticket attached by safety pin to main label at neck.
- Samples sent to the London office on Wednesday.

Docket 467 being knitted.

- All knitting done at home by 2/300 outworkers; factory inspects and packs.
- Ponchos knitted by hand, taking approximately 12 hours for one garment.
- Will London office please confirm to factory if cartons need plastic strapping to secure?

FACTORY J

- Jacket docket 4304.
- Fabric ordered waiting for confirmation of style change on pocket, will then make sample.
- Shipment for August 15.
- IJK 111 larger sizes; waiting for customer's comments.
- Docket 4229; quantity 2000 stock in production; first 100 washed and finished.
- Fabric felt too harsh. Factory rewashing and softening fabric and will make sure remainder are to correct standard. Measurements within tolerance.
- Sample sent to London on Wednesday.
- IJK 111 docket 4310 factory to counter sample lab dips. IJK 137 dockets 4311 and 4312 lab dips to be approved and then fabric to be ordered.
- Style IJK 111.
- Preproduction samples ready, but factory did not make ladies fastening, as requested.
- Measurements within tolerance.
- Brought samples back to take to customer and requested that factory make sizes 12 and 16 with correct fastening; should have been sent on seventh or eighth.
- Gave factory examples of mail order brand label and composition label and instructions such as do not remove tag jiffy label, bag and folding, and packing instructions.
- Delivery date to Manchester by sea on August 10; this is not possible, would have to leave this Saturday.

Factory J is an export company working with various factories in the area, but the jkt factory is its own and is next door to the its office.
- Approximately 7/800 workers making a variety of woven and fleece jkts.

SUMMARY

- All factories I visited were of a good standard and understand our critical path requirements, but it should be kept in mind that any of these factories could subcontract work out if it suited them.
- All companies will accept opening orders of approximately 1000 pieces. Factory G will accept minimum orders of 500 on some styles.
- Beware of YKK copies. I noticed in several factories that zips had YKK logo but were pin-locking zips, which YKK doesn't supply.
- All YKK zips should be auto-locking.
- All factories seem very keen to develop business with our companies.

I visited factories that were in production or just finished production, and they had picked out finished garments in a selection of sizes and colors ready for me to inspect, laid out in neat piles on a table in the comfortable air-conditioned showroom with a selection of cold drinks. My first reaction was how considerate, but then, I thought what if they had handpicked the best samples, checked them over very carefully, and repressed each garment before I looked at them, and would this really give an accurate assessment of the quality of the whole delivery? The answer of course is *no*.

14

Inspection of Merchandise

Before checking goods, always be prepared and have the approved sample, color swatches, specification, and all relevant information at hand. It makes your job much easier if you are always methodical, and this applies especially to inspection. You must be meticulous but also quick if you have many deliveries to check in a limited amount of time (Figure 14.1).

ACCEPTABLE QUALITY LEVELS

It is very unlikely that you will have a delivery with no faults, so you accept that a percentage of the production will have some major faults. The factory should always do a 100% inspection of the goods before packing and to what degree can vary from each factory (this is an important point to check when doing a factory assessment).

Often, it can just consist of trimming cotton ends and a brief glance at the garment. Other factories do a more detailed inspection. As the customer, you should not have to do a 100% inspection of every delivery, as it is very costly and time consuming, and it is the responsibility of the factory to check that the delivery is acceptable. However, it is essential that you do a check on the goods before accepting them.

Acceptable quality level (AQL) sampling plans are an internationally adopted method of determining the level of quality of a delivery by checking a percentage of the total quantity. A sample size of garments to inspect is determined by the quantity in the delivery and then the number of faulty garments you find after the inspection that you consider acceptable. If the number of faulty garments exceeds the acceptable level, then a 100%

FIGURE 14.1
Here is your order for 500,000 shirts.

inspection is required. The following tables are the most commonly used in the apparel industry (Figure 14.2).

The AQL inspection system was developed by the military during the World War II to check munitions and has now been adapted for deliveries of all types of products, as the level of rejection can be loosened or tightened, depending on the requirements of your business.

Here, we have an example of three levels of inspection used in the textile industry, and you can see that the 2.5 level allows less faulty garments than the 4.0 and 6.5 levels. You may decide to use 4.0 AQL for all your deliveries or, for example, use 2.5 AQL for more expensive luxury items.

It is important that your factories are aware of the level of inspection you will be using when checking deliveries at your warehouse or at their factory.

Ideally, the factory should do its AQL inspection on each delivery before it is shipped. It is common today for companies to use independent inspection companies or, if you have a local office, use your own representatives to do the AQL inspection at the factory before goods are shipped. If the delivery fails the inspection, it can then be rectified before shipping.

Whichever method or whomever we use for the inspection, our goal is ultimately to work with factories that build in quality checks at every stage of design and production, so that an AQL on their products becomes more routine than essential.

QUANTITY IN DELIVERY	SAMPLE SIZE	Acceptable quality level					
		2.5		4.0		6.5	
		Accept	Reject	Accept	Reject	Accept	Reject
2–8	2	0	1	0	1	0	1
9–15	3	0	1	0	1	0	1
16–25	5	0	1	0	1	0	1
26–50	8	0	1	1	2	1	2
51–90	13	1	2	1	2	2	3
91–150	20	1	2	2	3	3	4
151–280	32	2	3	3	4	5	6
281–500	50	3	4	5	6	7	8
501–1200	80	5	6	7	8	10	11
1201–3200	125	7	8	10	11	14	15
3201–10000	200	10	11	14	15	21	22
10001–35000	315	14	15	21	22	21	22

FIGURE 14.2
AQL inspection chart.

EXAMINING GARMENTS

A delivery of 3000 pairs of trousers would require 125 pairs to be inspected according to the AQL plan. The total number of cartons in the delivery would be approximately 100. You must be sure that the cartons you open are representative of all the colors and sizes in the delivery and your samples are taken randomly from carton 1 to 100. You are now checking a good representation of the delivery. When examining garments, it is important to keep to the same order of checking, following the same path.

- As you open the carton, check the box and bag quality.
- Check if the garment labels are correct.
- Check the fabric quality and color.
- Check the style.
- Check the measurements of one of each size of each color; if you are not happy with the measurements, check more as you go through the remainder of the sample batch.

At the same time when you are establishing that the above points are correct, check the workmanship and appearance. It's much easier to follow a predetermined path than letting your eyes wander randomly all over the garment. Inspect the back of the garment the same way as the front (Figure 14.3).

Seams may look okay, but a small amount of pressure should be applied to seams as you are checking the garments to make sure that they are securely sewn (Figure 14.4).

Inspection of Merchandise • 191

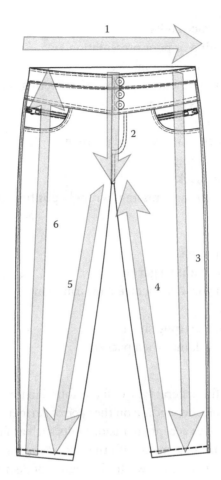

FIGURE 14.3
Inspection path.

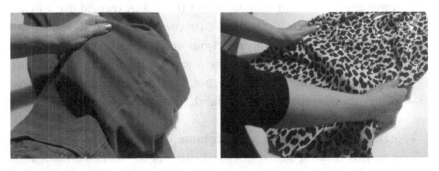

FIGURE 14.4
Inspection of seams.

- Examples of major faults.
- Wrong packaging or poor presentation.
- Wrong style or incorrect fabric.
- Incorrect measurements, significantly affecting the fit or style.
- Major fabric flaws or damages.
- Burst or split seams.
- Corners of pockets not secured properly.
- Broken top stitching.
- Stains or dirty marks.
- Buttons or fastenings missing, wrongly positioned, or not securely attached.
- Faulty zips.
- Badly fitting linings.
- Shaded panels within a garment.
- Labels missing or with incorrect information.
- Badly pressed.
- Garments have a strong smell.
- Garments were damp when packed.

I have categorized the general type of faults found in textiles, but there are, of course, many more, depending on the type of garment. Some faults that a factory would have to accept are major faults such as faulty zip, a button missing, a split seam, a garment with the wrong size label, but with many of the above faults, it comes down to the matter of degree of how bad the fault is. For example:

- Shading can occur if panels are cut from different rolls of fabric, or sometimes, color can change within the roll. A trained eye will see the color difference, but will the customer notice it?
- Often fabrics can have imperfections and the mills will say they are natural characteristics of the fabric, you say, but they're not showing on the sealed sample or they are, but not showing to the same degree as they show on the production fabric, but would the customer notice or think of it as a fault?
- Stains and dirty marks are a common problem with textiles, but even this is not a straightforward issue. Is the mark permanent or can it be easily brushed off? How big is the mark and where is it? How frequently do the marks appear in a noticeable area?

- A lining may fit badly, but does it affect the hang of the garment? Is it noticeable when worn?
- Poor pressing and poor presentation are definitely an issue if the garments will be hung on a rail in your store, but if the garment is packed in a bag and delivered to the customer, would the customer expect creasing and be quite prepared to press the garment before wearing?
- There are certain measurements that could be outside your acceptable tolerance but might not affect the fit of the garment. For example, the hem of a flared skirt may be 3" or 4" smaller than the specification. Does it dramatically affect the look of the skirt and would the customer be aware that the hem should have been that much larger?

There is also the risk of letting your own personnel's likes and dislikes influence your decision. One person might class a fault that you consider major as minor or not a fault at all, or what you consider minor, they would consider major. This can be a minefield and requires good commercial common sense. The selling price and end use of the garment will influence the final decision. What might be a minor fault in a cheap garment or possibly not considered fault could be a major fault in a more expensive garment. If there were a difference between the sealed sample and the delivery, would the customer realize? They don't know what you have bought; they will only judge by what they see on the rail. Even when buying from a photo from a catalog or from a Website, they are buying a concept rather than every detail of the garment. So, before making a final decision, reassess the faults, try garments on, and see how they look or, if possible, get other unbiased opinions.

One problem that regularly crops up is variations in the appearance of denim—the finished wash not being the same as the approval sample. There are several reasons for this: the take up of the dye is different and the production might have been washed at a slightly different temperature or not long enough. On my report from China, there were two problems that originated from the laundry: wrong color after washing and fabric that had not been softened enough during the wash cycle. If I had not been at the factory, these problems would not have been discovered until much later. There will always be inconsistencies with these washing processes, and we have to decide if the finished result is acceptable or too far removed from the approval sample.

If a delivery is rejected, it could mean cancellation, the cost of a 100% inspection, or other financial penalties for the supplier. With this in mind, a factory will try to make a case that the faults are not serious. Some manufacturers will say that if customers return the garments as faulty, they will take them back and give credit; this is an easy way out for them but could cause bad feeling between you and your customers.

Although you report back on the result of the inspection to the buyer, they will expect you to give a decision if the delivery is still fit to accept. If you say it is acceptable, then that is usually the end of it, but if you say it isn't, depending on the type of fault, they will want to see examples. They may disagree with you and say it is acceptable or may be willing to accept the faults in this instance, as commercial considerations may prevail (badly out of stock and losing money every day). In these cases, it would be the quality control manager or buying manager who makes the final decision. (Who makes the final decision will vary from company to company.) You may personally disagree with the decision, but you have done your job in highlighting a potential problem. Whether a delivery is accepted or not, this is not the end of the matter, because as part of your quality assurance procedures, you have to ensure that the problem does not occur again. We have to try and get to the root of the problem and first look at your own procedures and see if you overlooked anything in the writing of the specifications, approval of the samples, or control of the critical path, which could avoid the same problems happening again. A reoccurring problem with overseas factories is that samples might not be made in the factory but in a sample department. This is okay if the standard of the sample department is representative of the factory, but often, this is not the case, because the production is subcontracted to another factory. You need to know if this is happening, and your supplier should make it clear where your production is being made. Subcontracting is not a problem if the main factory carefully controls and supervises the subcontractor to the same degree as it supervises itself.

FACTORY HISTORY

Through the seasons, we receive many hundreds of deliveries from our vendors, and a record should be kept of how each factory performs. If we are having constant problems with a factory, such as deliveries that need reprocessing due to faulty packaging and labeling or 100% inspections due to the delivery failing the AQL inspection, this needs to be flagged up and a report should be compiled and sent to the buying department. There are suppliers who are very reliable and whose deliveries are always correct. We have a duty to make our company aware of these suppliers' good record and actively promote them to receive more business. Faulty deliveries have a knock-on effect, resulting in customer dissatisfaction, lost sales and profit, and wastage of time in sorting out the problems. Although we always aim for no faults, we accept that there might be a small percentage of faulty garments in a single delivery; at the same time, we also accept that a small percentage of the deliveries we receive might be rejected or need 100% inspection. In both the instances of the individual delivery and the multiple deliveries to your company, the priority is to keep faulty merchandise to a manageable amount. For single deliveries, we use the AQL table, and for multiple deliveries, we can use a similar guideline. I would aim for 5% or less faulty deliveries coming in to the warehouse or rejected at the factory if offshore auditors are employed. This means 5 out of 100 deliveries will be rejected or reprocessed. I'm not saying this is acceptable, but it is manageable, and from every delivery that is rejected, we should learn some lessons to prevent the problem from reoccurring.

15

More Preventative Action

The final chapter of this book deals with preventative action that can be taken at different stages to improve the products. Up to now, I have dealt with the day-to-day mechanics of quality assurance (QA), preventing problems with production.

To stay ahead of the game, we need to take a more in-depth critical view of what we sell and how we sell it. Fashion, particularly, is a capricious business, relying mainly on impulse buying, so we have to make it as problem free as possible for the customer. We cannot afford to sit back and be complacent, and the QA/technical department needs to be constantly re-assessing all aspects of the products we sell and to be checking if the QA function is working effectively.

A CASE STUDY - REVIEW OF A COMPANY'S QUALITY ASSURANCE

For a period, I worked as a freelancer in the men's and ladies' departments of a national mail order company, and I was asked to do a review of the QA department, and these were my findings.

Sizing and Grading

Inconsistencies in grading lengths, particularly on larger sizes, will affect sales. Standard grade rules need to be set for each category of merchandise, allowing the customers to buy the same size in different garments, knowing that they will fit. This is essential if sales in the larger-size market is to continue developing.

Quality Control Directives

The quality control manual makes brief references to packaging and labeling and does not go into sufficient detail. It should emphasize the importance of good overall presentation of the product, along with more detailed packaging instructions. During investigation of high-returning lines, I found items in too small a bag, on wrong-sized hangers, and incorrectly folded, causing unnecessary creasing and poor presentation, which gave the impression that the product was not of good quality and was poor value for money.

This does have a direct impact on returns, and I would question if there is sufficient feedback from the warehouse on deliveries that do not meet the agreed standard and if suppliers clearly understand what the requirements are and that they should be penalized if those standards are not met.

Supplier Manuals

At present, there are three separate manuals that cover quality issues:

1. Quality control manual
2. Textile performance standards
3. Packaging manual

I suggest it would be better to combine all three into a single comprehensive company manual. This could also include a letter of acceptance signed by a director of the company, agreeing to the terms and conditions of the QC procedures.

Factories are requested to inspect deliveries before shipping, but packaging and presentation are not referred to in the list of major faults. There should be a company standard inspection form for the factory to use as a checklist for all quality points, including packaging and presentation. It should also specify that the sample batch should be representative of all sizes and colors. Factories were sending approval samples to head office without any written report, and they should be instructed to complete a company sample report, with a checklist covering all aspects of the product.

Product Data Management

There are three software programmes for recording technical information. The technologist uses product data management (PDM) for specifications and a different one for reports; packaging has a separate programme. At present, buyers have no access or input into PDM, as was originally intended. The way the PDM is set up, details on the drawings of certain styles cannot be clearly shown. PDM should be re-assessed to decide if the company wants to continue with the system and to develop it to its full potential.

Summary

The company's approach to managing quality has varied and not been consistent. Several years before I joined them, they had a senior quality control manager, but at the time I started with the company, it did not have anyone in this position, and the responsibility for QA was divided between four buying managers. This meant that there was not a unified approach toward quality. Projects that had been started by the then quality manager had not been seen through to completion. There were approximately 10 experienced technologists working on textiles; their priorities were approval of samples and organization of fit sessions. This was a large company, selling a huge mix of products, and it needed a senior person who could bring all the strands of QA together. Money had been invested in software that analyzed returns, and every day, a report went to the technologists with a list of high-returning lines to investigate. With this and approving samples, the technical team had very little time to do anything about the issues I raised in my report.

This report was written more than 10 years ago, and I'm sure that this company has improved its methods, but quality assurance objectives are timeless and core values do not change only the the technology we use and how we apply it to the task at hand. Figure 15.1 shows a specification for a jean, using the PDM software, when I wrote the report. The technologist creates the drawing, but then he has to fit it into a predetermined box size. The technologist knows all the details of the product, but will the factory know the details when it receives this specification?

Several years ago, I worked for a national catalog company that did a major marketing campaign, selling men's and ladies' polo shirts. This became extremely successful owing to the space the company gave them in the catalog, and in each season, the company offered more colors, eventually offering up to twelve colors, in short and long sleeves, winter and summer weight fabrics, plus a children's range. The marketing strategy was that when the customers buy one color and like the garment, they will re-order from the other colors. Sales wildly exceeded expectations, and the buyers were desperately placing orders with numerous factories

SIZE CHART

Cat no Division Supplier Fabric	Technologist Buyer Date Approved

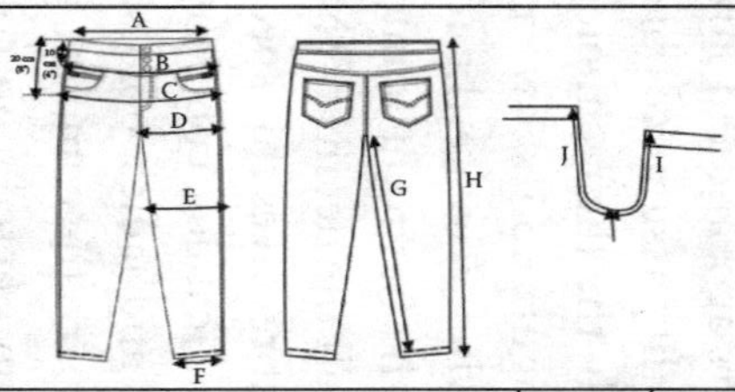

Ladies cotton trouser
Front zip twin stitched
Back and front yoke twin stitched
Zip pockets set in front curved side pockets
3 metal button front fastening
All seams 5 thread overlocked
Side seam, center back and center front seams twin stitched
Waist band twin stitched

SIZE	8	10	12	14	16	18	20	22		
A. WAIST OPENING	29¾"	31¾"	33⅞"	35¾"	37¾"	39¾"	41¾"	43¾"		
B. HIP 10 CM BELOW TOP OF THE WAISTBAND	34¼"	36¼"	38¼"	40¼"	42¼"	44	46	48½"		
C. HIP 20 CM BELOW TOP OF THE WAISTBAND	39⅜"	41⅜"	43⅜"	45⅜"	47⅜"	49⅜"	51⅜"	53⅜"		
D. THIGH AT CRUTCH	25⅜"	26⅜"	27½"	28¾"	29⅞"	31⅛"	32½"	34		
E. KNEE MIDPOINT BETWEEN HEM AND FORK	26"	26¾"	27½"	28⅜"	29⅛"	29⅞"	30⅝"	31½"		
F. HEM	10⅝"	11⅜"	12¼"	13	13¾"	14½"	15	15½"		
G. INSIDE LEG	29"	29"	29"	29"	29"	29"	29"	29"		
H. OUTSIDE LEG INCLUDING BAND	36¾"	37	37¼"	37½"	37¾"	38	38½"	39		
I. FRONT RISE, INCLUDING WAISTBAND	9	9⅜"	9¾"	10⅛"	10½"	11"	11½"	12		
J. BACK RISE, INCLUDING WAISTBAND	12¼"	12⅝"	13"	13⅜"	13⅞"	14¼"	14¾"	15¼"		
WAIST BAND DEPTH	1⅛"	1⅛"	1⅛"	1⅛"	1⅛"	1⅛"	1⅛"	1⅛"		
ZIP POCKET LENGTH	4⅜"	4⅜"	4⅜"	4⅜"	4⅜"	4⅜"	4⅜"	4⅜"		
BACK POCKET WIDTH	5⅛"	5⅛"	5⅛"	5⅛"	5⅛"	5⅛"	5⅛"	5⅛"		
BACK POCKET DEPTH	5⅛"	5⅛"	5⅛"	5⅛"	5⅛"	5⅛"	5⅛"	5⅛"		

FIGURE 15.1
Example of a company size chart.

worldwide. If a buyer has the best-selling lines in the catalog but no stock, it is akin to committing hara-kiri. The shirts from the original supplier were a good quality, and because the shirt was considered a basic garment, the buyers thought they could place orders with any company and the quality would be similar; unfortunately, it didn't work like that. The suppliers sensed that the buyers were desperate and they thought Christmas had come early. It became essential that I set some ground rules, because what would be the point of our customers being happy with their first buy but badly disappointed with the second and third garments. So, what standards did I insist on? The new suppliers had size charts, style description, and color swatches. But, what about the other details that would persuade the customers to buy more? Would the customer not expect each garment to be identical, except the color?

- Yarn to be combed or semi-combed to help make the fabric appearance satisfactory.
- Garments should have five thread over lock-secure seams.
- Placket opening to have interlining, especially to strengthen the button holes.
- Ribbed collar and cuffs to have elastane to keep their shape during wear and washing.
- Maximum shrinkage set, so the garment size does not alter noticeably after washing.
- Set a standard for dye fastness to washing and wet and dry rub, so that color does not fade or dye does not stains other garment either in the wash or when being worn.

These polo shirts are not throw-away fashion items. The company's customers were generally an older age group, who expected a quality garment, which would keep its shape and color after many washes, and this would be the influencing factor to encourage them to buy more items. When setting quality standards, it is important to know your customer, and I convinced the buying department that it was commercial sense to insist on these standards and that it was a commercial suicide if we didn't. Another aspect to consider is if new customers are attracted to buy the polo shirts and they are happy with the quality, they will stay as customers and buy other merchandise. There are many reasons why customers stay loyal to a brand. Quality or the customer's perception of quality is usually one of those reasons, and different customer profiles have different perceptions of quality, and knowing what motivates your customer is the key to success.

INVITING FACTORY PERSONNEL TO YOUR HEAD OFFICE

Over the years, many factory owners with their merchandisers have visited the companies that I worked for.

This is an excellent opportunity to explain how you work and show off your company. Grasp this chance to train your suppliers in working according to your methods and see the business through your eyes. Meeting personally face to face at the factory or your head office helps strengthen the relationship between your two companies. There will be a queue of people wanting to meet the visitors such as buying and merchandising staff, the shipping department, warehousing, finance, and operations managers, who they communicate with all the year round, and, through these face-to-face meetings, wanting to resolve many problems and strengthen working relationships.

FAULTY RETURNS

Every company has a corner in the warehouse where faulty garments returned from customers are stored. They will be written off or jobbed off, but before doing that, the quality controller should take time to look through these returned items. You will take on the role of detective, tracking down serial offenders. If the returns department is managed properly, they will contact you if there is an abnormal amount of returns for a particular fault or if a particular style has a high return rate. These returns are often the first indication that you have a problem with a line and this might then necessitate checking the stock in the warehouse. No one wants faulty goods to go out to customers, but in some instances, even after all the checks that you have set in place, faults might not show themselves until after the garments are worn, so looking at returns can be a very productive procedure, which can alert you to potential future problems.

When I first joined a mail order company, I was shown around the warehouse and my visit to the returns department coincided with the morning's post; thousands of items were emptied from sacks onto an enormous carousel to be reprocessed and returned to stock. It's a fact that the same garments can go out to customers and returned to the warehouse several times. All returns should be checked carefully to break the cycle of continuous returns, which are very expensive to process. QA can play a role in ensuring that goods are renovated and packed as originally specified, especially when doing stock checks and reporting any problems back to the warehouse.

SCARY MOMENTS

Sometimes, you have instances that for a quality controller are downright scary. When working for an importer, we were approached by a catalog company and asked how quickly we could supply a padded jacket that has been selling in their winter range; the original supplier could not produce any more in time to satisfy the demand to the end of the catalog, so the company approached us, as we had a reputation for doing things quickly and giving good service. There was one stipulation they had to be made in the UK that they could not take the chance of delays due to shipping and quality if sourced from offshore factories. We tried very hard to get them made in the UK; one factory started to make them, but it was very slow. The directors had as a backup asked one of our regular suppliers in China to sample the jacket, and when they sent an acceptable sample, they decided to place the order in China. With the cooperation of the factory, the goods were shipped on time and the quality was correct. The customer was very happy with the goods, never knowing where they really originated from and still believing that they were made 100 miles down the road. (Country of origin label was not a requirement at that time.)

The same company was asked to supply a pack of two value trousers. These were ladies' knitted polyester trousers with an elasticated waist, and a pack of two consisted of two different colors. They were offered in several leg lengths. Originally, I think they forecast to sell 20,000–30,000 packs. Sales hit the roof and they finished up selling more than 100,000 packs. Their merchandiser took over an office in our company, and we were getting large deliveries nearly every day. To readers in the States, these quantities might seem small, but the trousers were advertised in a flyer sent through the post and the lead times very short; quantities were hard to estimate until the orders came in, as there was no previous history for this product. To take on this type of project, you need a very good supply chain and, most importantly, factories that you can rely on 100%, and thankfully, we had both. It succeeded because both the importer and the factory wanted to increase business with this customer; we needed each other, which is always a good formula for success.

A TRIP TO THE STORES

If you work for a group of stores or shops, take time to talk to the sales associates on the front line who deal directly with the customers. They will be pleased that someone from head office is willing to listen to their opinion of the merchandise, and much useful information can come out of this. For example, size and fit are often finalized after trying sample garments on the girl in the office, who, it has been decided, is a standard size 12 or 14. This often works very well, but if the majority of customers are not buying a dress, because they complain that the front armhole seam cuts into them and the armhole is too small, who do you take notice of? If there are more orders in the pipeline, you might be able to alter the pattern in time, and sales will increase when the new amended production comes into the store. Warning: do not take everything you are told on the shop floor as gospel; any problems that you are told about need to be verified and agreed between you and the buyer first, but this can be a good source of information.

BENCHMARKING

This term is used when comparing the quality of your merchandise with that of your competitors, and it can be really valid only if you are comparing like for like. This is only worth doing for ranges that you sell season after season and that change very little. For example, school wear, budget denim ranges, classic men's shirts and ladies' blouses, and budget trousers and skirts. There are many more examples, but the criterion is the garments that your company and your competitors sell in large quantities season after season to a similar target customer. The comparisons should be for all of the following:

- Fabric quality
- Sizing
- Workmanship
- Packaging and presentation

This is a good method of re-evaluating your core lines, especially if you feel that returns are high or if you are experiencing a higher percentage of customer complaints than normal. Take the opportunity to benchmark against at least four or five competitors' merchandise; the more, the better, so that you can learn more about the differences between what you are selling and what other retailers are offering.

TEAM BUILDING

This is a term often used today, usually associated with company's employees spending a day together, in which they form teams and are given tasks to do, which are totally unconnected to their everyday work. The events can be themed such as taking on the role of detectives to solve a murder; making a skyscraper from bits of cardboard, glue, and staples; and other mental and physical challenges. The aim is to take people out of their everyday environment, have fun bonding, which will result in them working better together when they go back to the office the next day. I have been on one of these events, and as the day progressed, I became more pre-occupied with how I was going to catch up next day with the work accumulating in my office.

Building a team that works together effectively is crucial in any business and especially so in fashion. I have worked in many different set-ups: quality control and buying being in the same building but on different floors and even in different towns. The set-up that I find the most effective is the buyer, merchandiser, and QA working together in the same office. Lines of division begin to blur as each member of the team takes on an appreciation of the other team member's problems and responsibilities. There should be a continuous flow of information, discussion, and exchange of ideas, and each team member should feel that he or she contributes equally to the department's efforts and successes. Most importantly, quality problems can be resolved more quickly and efficiently within the team, as the other members of the group see it as their problem as well.

ODD BEDFELLOWS

In the final section of this book, I wanted to recall a couple of instances of strange partnerships between buying companies and suppliers, which had a direct impact on quality.

A catalog company that I worked for many years ago sold a range of budget blazers in 100% polyester for men, and the same style was modified for school wear. Over the years, they sold several thousands, and to keep the price very competitive, they sourced the blazer from an agent in Korea, hoping that this would eventually lead to increased sales. At the earliest opportunity, I visited the factory with the agent Mr. Pi. The factory was the biggest that I had ever seen, employing thousands of machinists, organized into dozens of production lines. The factory manager showed me every department and then stopped in front of a production line and proudly told me that this line of machinists was to be set aside exclusively for our orders. I'm no mathematician, but it did not take long for me to work out that given the number of garments this line could produce in a week, our entire year's production could be made in 2 weeks. I realized Mr. Pi was an extreme optimist; he knew our business, and had he led the factory manager on a little? The factory produced very-good-quality items, but it had been set up for the American market and talked in telephone numbers. The factory believed our first order was a trial order, and when this was produced to our satisfaction, we would start to place the real quantities. Unfortunately, I never visited that factory again, but every year, Mr. Pi would find another factory, and the process would start all over again, never achieving continuity of quality.

The second instance was a manufacturing company that had moved most of its production offshore to a factory in Eastern Europe. Its prices were still too high, and it was losing business to competitors that made similar garments in China. It chose to work with a manufacturer in China, who had its own UK office approximately 10 miles from its company and who also employed a quality controller based at this local office. The China supplier was a big concern with many resources; what could go wrong? Well, many things did go wrong, and at the beginning, it was hard to explain why. The MD of the manufacturer's local office with the quality controller visited their customers on many occasions, but problems, which should have never happened, continued. One day, I saw an article about this supplier; it was one of the biggest in China and the UK business was approximately 5% of its total business. So, what percentage was this company's business of that 5%? May be, 0.00001%, at a guess. This explained why it was having problems; no manufacturer will turn down business, but if it is insignificant to them, you have to trade under their terms and conditions; they will not take the time to understand the details of your business and what is important to you. As far as the quality of the products is concerned, it will have no impact on them if they lose the business.

Summary

In addition to discussing the mechanics of quality assurance and the tools we use, I have also devoted time to discuss the dynamics at play in any given scenario. It is very important that a quality controller demonstrates his or her integrity in any situation as you take on the role of a "policeman" and may require the judgment of Solomon. Wow you might be thinking that this all sounds a bit heavy! While I am at it, you also have to be a skilled diplomat. I wanted to prove that the quality assurance is not a passive discipline, just ticking boxes, but should be very proactive and a very high-profile role. Most of the people in your company will admit that quality is very important to the business but only the quality controller will have the commitment, will, and the ability to see where things need to be changed. To me this is the essence of the work, and to succeed in any job you have to appreciate its real contribution and potential for improving the business that you are working in.

I have worked with quality controllers and garment inspectors who are triumphant when they find what they consider to be a faulty delivery. They view it as justifying the job they do and the salary they earn, they have caught someone out trying to sneak in poor quality through the backdoor, but not on their watch, nothing can get past them! However when a senior manager looks at the results of the inspection and decides that it is commercially acceptable, they shake their heads in disbelief and wonder why they bother inspecting anything at all. When I first started this job I reacted in the same way, but quickly I realized that if this was my business I would be disappointed if I found faulty goods as this would mean losing money if they were rejected. However, if after carefully assessing the problems, the delivery was considered commercially acceptable, I would be relieved that the goods were still saleable. Throughout this book, I give many examples of the importance of being able to make good commercial decisions.

Quality assurance is a complicated subject, and I hope that I have addressed many of the issues and problems you will come across and my intention is to give you sufficient information so that you have a clear picture of what the job involves and when you start in this challenging career you will be able to "hit the ground running."

Index

Note: Page numbers followed by f and t refer to figures and tables, respectively.

A

Abrasion test, 98
Acceptable quality level (AQL), 187–189
    inspection chart, 189f
Amendments to
    sweatshirt, 170, 172f
    trouser, 170, 173f
Anoraks, 24f
AQL. *See* Acceptable quality level (AQL)

B

Back and cuff, 28f
Back and strap, 58f
Bag(s)
    for briefs, 134f
    with cardboard inserts, 135f
    measurements of, 57f
Benchmarking, 204–205
Buttons, 32f
    fastening and braid, 71f
Buying department, 3, 8, 164, 195, 201

C

CAD, 18
Candidate profile, job brief, 5
Care label position, 145f
Carton, 190
    specification, 147
Chain of departmental stores, catalog companies, and internet shops, 2–3
Cleanliness, 177
Collar, 78f
    construction, 29f
    and placket, 40f
    styles, 117f

Color fastness test, 98–99
Company profiles, 2–4
    chain of departmental stores, catalog companies, and internet shops, 2–3
    import agents, 3
    international high fashion brand with shops and in-store concessions, 2
    new label, 3–4
Company's bible, 101
Company's QA, 197–202
    PDM, 198–199
    QC directives, 198
    sizing and grading, 197
    summary, 199–202, 200f
    supplier manuals, 198
Company standard inspection, 198
Compartments, 61f, 62f
Core basic lines, size charts and specifications, 108–124
CorelDraw, 18
Core lines/styles, 20
Counter samples, 12
Critical path, 11, 13
    managing, 163–167, 165f
Cuff styles, 116f

D

Delivery
    inspections, 3
    of material to factory, 164
    to stores, 163
    to warehouse, 163
Denim, 193

E

Examination of garments, 190–194, 191f

211

F

Fabric(s), 3, 181
  department, 2
  minimum performance standards, 103–104, 105t, 106t, 107t
  new, basic properties, 99
  performance specification, 93
  testing, 97–99
    abrasion and pilling, 98
    color fastness, 98–99
    construction, 97
    seam slippage, 98
    seam strength, 98
    stability to washing and dry cleaning, 98
    tear strength, 98
    weight, 97–98
Factory history, 194–195
Factory management, 8, 176
Fashion brands, 19
Fashion wool jacket, 94
Faulty returns, 202–203
Fit and fit sessions, 89–91
Flat measurement, 19–20
Flat packed garments, 147
Front logo, 59f

G

Garment
  balance, 89
  interlinings, 26f
  packaging, 35f
  size chart, 33f, 42f, 52f, 66f, 73f, 80f, 86f
  summer, 36
Gilet
  folding and bag instructions for, 146f
  labeling, 143f
  label positions at neck for, 144f
Girls' dress
  development, 155–161, 156f
    construction details, 157f
    dress in hanging bag, 161f
    labels and hangers, 158f
    measuring points, 159f
    size chart, 160f
  with full skirt, 68–74, 68f, 70f
Golden rule, 15

H

Heading test, 103

I

Import agents, 3
In-house models, 90
Inner waistband, 50f
Inspection
  AQL, 187–189, 189f
  levels, 189
  merchandise, 187–195, 188f
International high fashion brand with shops and in-store concessions, 2
Intertek
  jersey standards, 105t
  knitwear standards, 106t
  woven standards, 107t
Inviting factory personnel to head office, 202
ISK 326 black mini with diamante's, 180
ITR 318, 180
  bigger sizes, 180–181

J

Jeans
  display on hanger and labeling, 129f
  on hanger, 130f
  range, 149–155
    back with hip pockets, 155f
    front pocket, zip and belt loop measurements, 154f
    leg shapes, 151f
    measuring points, 153f
    size chart, 152f
Job brief, 4–5
  candidate profile, 5

K

Knitted jersey top, 38f
Knock-on effect, 195

L

Label and embroidery, 41f
Label position, 118f

Ladies
  basic
    line skirt, 124f
    shirt, 123f
  briefs, pack of, 133f
Ladies' fashion trouser, 45–53, 45f, 47f, 48f
Ladies' tie dress with bodice seams, 82–87, 82f, 84f
Ladies' unlined motorbike jacket, 75–81, 75f, 77f
Laundering process, 93
Lining
  construction, 72f
  individual components, 31f
  inside, 30f

M

Manchester office, 178
Mannequin, 90
Manufacturing
  guidelines, 125
  time, 164
Marketing strategy, 199
Measuring points
  girls' dress, 159f
    with full skirt, 74f
  jeans, 153f
  ladies basic
    line skirt, 124f
    shirt, 123f
  ladies' fashion trouser, 51f
  ladies' tie dress with bodice seams, 87f
  ladies' unlined motorbike jacket, 81f
  men's cardigan, 67f
  men's short-sleeve knitted shirt, 43f
  reefer coat, 34f
  of shirts, 121f
Men's cardigan, 63–67, 63f
Men's short-sleeve knitted shirt, 36–44, 36f, 39f
Merchandise, 178
  inspection, 187–195, 188f
    AQL, 187–189
    examining garments, 190–194
    factory history, 194–195
Mock leather, 78f
Munitions, 189
Murphy's law, 18

N

Natural fibers, 93

O

Odd bedfellows, 206–207

P

Packaging instructions, 44f, 53f
Pattern cutter, 20, 176
Pilling test, 98
Planning production, 166
Plastic single butterfly, 140f
Pocket, 27f
  cuff and waist band, 79f
  styles, 115f
Police tunics, 18–19
Poly/cotton blend, 94
Product data management (PDM), 198–199
Product development, 149
Production samples, 163–164
Prototype, 164

Q

QA. *See* Quality assurance (QA)
QC. *See* Quality control (QC)
Quality
  audit's responsibility, 1
  controller, 169, 176, 178
    fit session, 89
  controller's responsibility, 13
  issues, 7
Quality assurance (QA), 1, 9, 13, 178
  aim of, 13
Quality control (QC), 2
  directives, 198
  role, 1–2
  team, 2
Quilted coats, 24f

R

Reefer coat, 22–35, 22f
Report from China, 177–185, 179f
  factory A, 179–180
  factory B, 181

Report from China (*Continued*)
  factory C, 181
  factory D, 182
  factory E, 182
  factory F, 182–183
  factory G, 183
  factory H, 183
  factory I, 183–184
  factory J, 184–185
Ribs, 65f
Risk analysis, 7–9
Rucksack, 54–62, 54f, 56f

S

Sample/delivery report, 171f
Sample department, 194
Sampling and production process, 13, 14f
Sampling procedures, supplier's manual, 102–103
Sand blasting operation, 178, 179f
Satin cotton, 181
Scary moments, 203–204
Sealing samples/initial samples, 164
Seam slippage test, 98
Seam strength test, 98
Seven belt loops, 7
Shipping time, 163
Shirts, 201
  folded into a bag, 141f
  measuring points, 121f
  packaging, 140f
Shower-resistant garments, 24f
Side pouch pockets, 60f
Six belt loops, 7
Size chart measurement, 19
Sizing and grading, 197
Socks
  cardboard packaging for, 138f
  pack of five, 137f
  steps to fold and positioned in package, 139f
Specification, 17–88
  core basic lines, 108–124
  girls' dress with full skirt, 68–74, 68f, 70f
  ladies' fashion trouser, 45–53, 45f, 47f, 48f
  ladies' tie dress with bodice seams, 82–87, 82f, 84f
  ladies' unlined motorbike jacket, 75–81, 75f, 77f
  men's cardigan, 63–67, 63f
  men's short-sleeve knitted shirt, 36–44, 36f, 39f
  reefer coat, 22–35, 22f
  rucksack, 54–62, 54f, 56f
  sending, 88
  standard carton, 147, 147f
Standard carton specification, 147, 147f
Stepping stone, 13
Style C1 standard fit casual shirt long sleeve, 112f
Style C2 tapered casual shirt long sleeve, 113f
Style C3 standard fit casual shirt short sleeve, 114f
Style F1 standard-fit shirt, long sleeve, 109f
Style F2 tapered fit shirt long sleeve, 110f
Style F3 standard fit shirt short sleeve, 111f
Subcontracting, 194
Summer garment, 36
Supplier's manual, 198
  basic size charts and specifications for core lines, 108–124
  fabric minimum performance standards, 103–104, 105t, 106t, 107t
  manufacturing guidelines, 125
  packaging and presentation, 125–130
  sampling procedures, 102–103
  standard carton specification, 147, 147f
  trouser hanger with bar, 130, 131f
  trouser's inside, 131–146
Supply chain, 11–12
  writing procedures, 13–15, 14f
Sweatshirt
  amendments, 170, 172f
  style, 8

T

Team building, 205–206
Technical designer, 169
Technical pack, 20
Technical responsibility, 1
Trip to stores, 204
Trouser
    amendments, 170, 173f
    hanger
        with bar, 130, 131f
        and labels, 128f
        inside with label position, 131–146, 132f
    manufacturer, 94

W

Warehouse, 202–203
Woven cloth factory, 181

Y

Yarn, 63
    count, 95
YKK zips, 180

Z

Zip, 8–9, 154f